Biology: Exploring Life

SECOND EDITION

Biology: Exploring Life

Gil Brum
California State Polytechnic University, Pomona

Larry McKane
California State Polytechnic University, Pomona

Gerry Karp
Formerly of the University of Florida, Gainesville

JOHN WILEY & SONS, INC.
New York / Chichester / Brisbane / Toronto / Singapore

Acquisitions Editor *Sally Cheney*
Developmental Editor *Rachel Nelson*
Marketing Manager *Catherine Faduska*
Associate Marketing Manager *Deb Benson*
Senior Production Editor *Katharine Rubin*
Text Designer/"Steps" Illustration Art Direction *Karin Gerdes Kincheloe*
Manufacturing Manager *Andrea Price*
Photo Department Director *Stella Kupferberg*
Senior Illustration Coordinator *Edward Starr*
"Steps to Discovery" Art Illustrator *Carlyn Iverson*
Cover Design *Meryl Levavi*
Text Illustrations *Network Graphics/Blaize Zito Associates, Inc.*
Photo Editor/Cover Photography Direction *Charles Hamilton*
Photo Researchers *Hilary Newman, Pat Cadley, Lana Berkovitz*
Cover Photo *James H. Carmichael, Jr./The Image Bank*

This book was set in New Caledonia by Progressive Typographers and printed and bound by Von Hoffmann Press.
The cover was printed by Lehigh. Color separations by Progressive Typographers and Color Associates, Inc.

Recognizing the importance of preserving what has been written, it is a policy of
John Wiley & Sons, Inc. to have books of enduring value published in the
United States printed on acid-free paper, and we exert our best efforts to that end.

The paper in this book was manufactured by a mill whose forest management
programs include sustained yield harvesting of its timberlands. Sustained yield
harvesting principles ensure that the number of trees cut each year does not
exceed the amount of new growth.

Library of Congress Cataloging in Publication Data:
Brum, Gilbert D.
Biology : exploring life/Gil Brum, Larry McKane, Gerry Karp.—2nd ed.
p. cm.
Includes bibliographical references and index.
ISBN 0-471-54408-6 (cloth)
1. Biology. I. McKane, Larry. II. Karp, Gerald. III. Title.
QH308.2.B78 1993
574—dc20 93-23383
CIP

Unit I ISBN 0-471-01827-9 (pbk)
Unit II ISBN 0-471-01831-7 (pbk)
Unit III ISBN 0-471-01830-9 (pbk)
Unit IV ISBN 0-471-01829-5 (pbk)
Unit V ISBN 0-471-01828-7 (pbk)
Unit VI ISBN 0-471-01832-5 (pbk)

Printed in the United States of America

10 9 8 7 6 5 4 3 2 1

For the Student, we hope this book helps you discover the thrill of exploring life and helps you recognize the important role biology plays in your everyday life.

To Margaret, Jan, and Patsy, who kept loving us even when we were at our most unlovable.

To our children, Jennifer, Julia, Christopher, and Matthew, whose fascination with exploring life inspires us all. And especially to Jenny—we all wish you were here to share the excitement of this special time in life.

Preface to the Instructor

Biology: Exploring Life, Second Edition is devoted to the process of investigation and discovery. The challenge and thrill of understanding how nature works ignites biologists' quests for knowledge and instills a desire to share their insights and discoveries. The satisfactions of knowing that the principles of nature can be understood and sharing this knowledge are why we teach. These are also the reasons why we created this book.

Capturing and holding student interest challenges even the best of teachers. To help meet this challenge, we have endeavored to create a book that makes biology relevant and appealing, that reveals biology as a dynamic process of exploration and discovery, and that emphasizes the widening influence of biologists in shaping and protecting our world and in helping secure our futures. We direct the reader's attention toward principles and concepts to dispel the misconception of many undergraduates that biology is nothing more than a very long list of facts and jargon. Facts and principles form the core of the course, but we have attempted to show the *significance* of each fact and principle and to reveal the important role biology plays in modern society.

From our own experiences in the introductory biology classroom, we have discovered that

- emphasizing principles, applications, and scientific exploration invigorates the teaching and learning process of biology and helps students make the significant connections needed for full understanding and appreciation of the importance of biology; and
- students learn more if a book is devoted to telling the story of biology rather than a recitation of facts and details.

Guided by these insights, we have tried to create a process-oriented book that still retains the facts, structures, and terminology needed for a fundamental understanding of biology. With these goals in mind, we have interwoven into the text

1. an emphasis on the ways that science works,
2. the underlying adventure of exploration,
3. five fundamental biological themes, and
4. balanced attention to the human perspective.

This book should challenge your students to think critically, to formulate their own hypotheses as possible explanations to unanswered questions, and to apply the approaches learned in the study of biology to understanding (and perhaps helping to solve) the serious problems that affect every person, indeed every organism, on this planet.

THE DEVELOPMENT STORY

The second edition of *Biology: Exploring Life* builds effectively on the strengths of the First Edition by Gil Brum and Larry McKane. For this edition, we added a third author, Gerry Karp, a cell and molecular biologist. Our complementary areas of expertise (genetics, zoology, botany, ecology, microbiology, and cell and molecular biology) as well as awards for teaching and writing have helped us form a balanced team. Together, we exhaustively revised and refined each chapter until all three of us, each with our different likes and dislikes, sincerely believed in the result. What evolved from this process was a satisfying synergism and a close friendship.

THE APPROACH

The elements of this new approach are described in the upcoming section "To the Student: A User's Guide." These pedagogical features are embedded in a book that is written in an informal, accessible style that invites the reader to explore the process of biology. In addition, we have tried to keep the narrative focused on *processes*, rather than on static facts, while creating an underlying foundation that helps students make the connections needed to tie together the information into a greater understanding than that which comes from memorizing facts alone. One way to help students make these connections is to relate the fundamentals of biology to humans, revealing the human perspective in each biological principle, from biochemicals to ecosystems. With each such insight, students take a substantial step toward becoming the informed citizens that make up responsible voting public.

We hope that, through this textbook, we can become partners with the instructor and the student. The biology teacher's greatest asset is the basic desire of students to understand themselves and the world around them. Unfortunately, many students have grown detached from this natural curiosity. Our overriding objective in creating this book was to arouse the students' fascination with exploring life, building knowledge and insight that will enable them to make real-life judgments as modern biology takes on greater significance in everyday life.

THE ART PROGRAM

The diligence and refinement that went into creating the text of *Biology: Exploring Life,* Second Edition characterizes the art program as well. Each photo was picked specifically for its relevance to the topic at hand and for its aesthetic and instructive value in illustrating the narrative concepts. The illustrations were carefully crafted under the guidance of the authors for accuracy and utility as well as aesthetics. The value of illustrations cannot be overlooked in a discipline as filled with images and processes as biology. Through the use of cell icons, labeled illustrations of pathways and processes, and detailed legends, the student is taken through the world of biology, from its microscopic chemical components to the macroscopic organisms and the environments that they inhabit.

SUPPLEMENTARY MATERIALS

In our continuing effort to meet all of your individual needs, Wiley is pleased to offer the various topics covered in this text in customized paperback "splits." For more details, please contact your local Wiley sales representative. We have also developed an integrated supplements package that helps the instructor bring the study of biology to life in the classroom and that will maximize the students' use and understanding of the text.

The *Instructor's Manual,* developed by Michael Leboffe and Gary Wisehart of San Diego City College, contains lecture outlines, transparency references, suggested lecture activities, sample concept maps, section concept map masters (to be used as overhead transparencies), and answers to study guide questions.

Gary Wisehart and Mark Mandell developed the test bank, which consists of four types of questions: fill-in questions, matching questions, multiple-choice questions, and critical thinking questions. A computerized test bank is also available.

A comprehensive visual ancillary package includes four-color transparencies (200 figures from the text), *Process of Science* transparency overlays that break down various biological processes into progressive steps, a video library consisting of tapes from Coronet MTI, and the *Bio Sci* videodisk series from Videodiscovery, covering topics in biochemistry, botany, vertebrate biology, reproduction, ecology, animal behavior, and genetics. Suggestions for integrating the videodisk material in your classroom discussions are available in the instructor's manual.

A comprehensive study guide and lab manual are also available and are described in more detail in the User's Guide section of the preface.

Acknowledgments

It was a delight to work with so many creative individuals whose inspiration, artistry, and vital steam guided this complex project to completion. We wish we were able to acknowledge each of them here, for not only did they meet nearly impossible deadlines, but each willingly poured their heart and soul into this text. The book you now hold in your hands is in large part a tribute to their talent and dedication.

There is one individual whose unique talent, quick intellect, charm, and knowledge not only helped to make this book a reality, but who herself made an enormous contribution to the content and pedagogical strength of this book. We are proud to call Sally Cheney, our biology editor, a colleague. Her powerful belief in this textbook's new approaches to teaching biology helped instill enthusiasm and confidence in everyone who worked on it. Indeed, Sally is truly a force of positive change in college textbook publishing–she has an uncommon ability to think both like a biologist and an editor; she knows what biologists want and need in their classes and is dedicated to delivering it; she recognizes that the future of biology education is more than just publishing another look-alike text; and she is knowledgeable and persuasive enough to convince publishers to stick their necks out a little further for the good of educational advancement. Without Sally, this text would have fallen short of our goal. With Sally, it became even more than we envisioned.

Another individual also helped make this a truly special book, as well as made the many long hours of work so delightful. Stella Kupferberg, we treasure your friendship, applaud your exceptional talent, and salute your high standards. Stella also provided us with two other important assets, Charles Hamilton and Hilary Newman. Stella and Charles tirelessly applied their skill, and artistry to get us images of incomparable effectiveness and beauty, and Hilary's diligent handling helped to insure there were no oversights.

Our thanks to Rachel Nelson for her meticulous editing, for maintaining consistency between sometimes dissimilar writing styles of three authors, and for keeping track of an incalculable number of publishing and biological details; to Katharine Rubin for expertly and gently guiding this project through the myriad levels of production, and for putting up with three such demanding authors; to Karin Kincheloe for a stunningly beautiful design; to Ishaya Monokoff and Ed Starr for orchestrating a brilliant art program; to Network Graphics, especially John Smith and John Hargraves, who executed our illustrations with beauty and style without diluting their conceptual strength or pedagogy, and to Carlyn Iverson, whose artistic talent helped us visually distill our "Steps to Discovery" episodes into images that bring the process of science to life.

We would also like to thank Cathy Faduska and Alida Setford, their creative flair helped us to tell the story behind this book, as well as helped us convey what we tried to accomplish. And to Herb Brown, thank you for your initial confidence and continued support. A very special thank you to Deb Benson, our marketing manager. What a joy to work with you, Deb, your energy, enthusiasm, confidence, and pleasant personality bolstered even our spirits.

We wish to acknowledge Diana Lipscomb of George Washington University for her invaluable contributions to the evolution chapters, and Judy Goodenough of the University of Massachusetts, Amherst, for contributing an outstanding chapter on Animal Behavior.

To the reviewers and instructors who used the First Edition, your insightful feedback helped us forge the foundation for this new edition. To the reviewers, and workshop and conference participants for the Second Edition, thank you for your careful guidance and for caring so much about your students.

Dennis Anderson, *Oklahoma City Community College*
Sarah Barlow, *Middle Tennessee State University*
Robert Beckman, *North Carolina State University*
Timothy Bell, *Chicago State University*
David F. Blaydes, *West Virginia University*
Richard Bliss, *Yuba College*
Richard Boohar, *University of Nebraska, Lincoln*
Clyde Bottrell, *Tarrant County Junior College*
J. D. Brammer, *North Dakota State University*
Peggy Branstrator, *Indiana University, East*
Allyn Bregman, *SUNY, New Paltz*
Daniel Brooks, *University of Toronto*
Gary Brusca, *Humboldt State University*
Jack Bruk, *California State University, Fullerton*
Marvin Cantor, *California State University, Northridge*
Richard Cheney, *Christopher Newport College*
Larry Cohen, *California State University, San Marcos*
David Cotter, *Georgia College*
Robert Creek, *Eastern Kentucky University*
Ken Curry, *University of Southern Mississippi*
Judy Davis, *Eastern Michigan University*
Loren Denny, *southwest Missouri State University*
Captain Donald Diesel, *U. S. Air Force Academy*
Tom Dickinson, *University College of the Cariboo*

Mike Donovan, *Southern Utah State College*
Robert Ebert, *Palomar College*
Thomas Emmel, *University of Florida*
Joseph Faryniarz, *Mattatuck Community College*
Alan Feduccia, *University of North Carolina, Chapel Hill*
Eugene Ferri, *Bucks County Community College*
Victor Fet, *Loyola University, New Orleans*
David Fox, *Loyola University, New Orleans*
Mary Forrest, *Okanagan University College*
Michael Gains, *University of Kansas*
S. K. Gangwere, *Wayne State University*
Dennis George, *Johnson County Community College*
Bill Glider, *University of Nebraska*
Paul Goldstein, *University of North Carolina, Charlotte*
Judy Goodenough, *University of Massachusetts, Amherst*
Nathaniel Grant, *Southern Carolina State College*
Mel Green, *University of California, San Diego*
Dana Griffin, *Florida State University*
Barbara L. Haas, *Loyola University of Chicago*
Richard Haas, *California State University, Fresno*
Fredrick Hagerman, *Ohio State University*
Tom Haresign, *Long Island University, Southampton*
Jane Noble-Harvey, *University of Delaware*
W. R. Hawkins, *Mt. San Antonio College*
Vernon Hedricks, *Brevard Community College*
Paul Hertz, *Barnard College*
Howard Hetzle, *Illinois State University*
Ronald K. Hodgson, *Central Michigan University*
W. G. Hopkins, *University of Western Ontario*
Thomas Hutto, *West Virginia State College*
Duane Jeffrey, *Brigham Young University*
John Jenkins, *Swarthmore College*
Claudia Jones, *University of Pittsburgh*
R. David Jones, *Adelphi University*
J. Michael Jones, *Culver Stockton College*
Gene Kalland, *California State University, Dominiquez Hills*
Arnold Karpoff, *University of Louisville*
Judith Kelly, *Henry Ford Community College*
Richard Kelly, *SUNY, Albany*
Richard Kelly, *University of Western Florida*
Dale Kennedy, *Kansas State University*
Mirium Kittrell, *Kingsboro Community College*
John Kmeltz, *Kean College New Jersey*
Robert Krasner, *Providence College*
Susan Landesman, *Evergreen State College*
Anton Lawson, *Arizona State University*
Lawrence Levine, *Wayne State University*
Jerri Lindsey, *Tarrant County Junior College*
Diana Lipscomb, *George Washington University*
James Luken, *Northern Kentucky University*
Ted Maguder, *University of Hartford*
Jon Maki, *Eastern Kentucky University*
Charles Mallery, *University of Miami*
William McEowen, *Mesa Community College*
Roger Milkman, *University of Iowa*
Helen Miller, *Oklahoma State University*
Elizabeth Moore, *Glassboro State College*
Janice Moore, *Colorado State University*
Eston Morrison, *Tarleton State University*
John Mutchmor, *Iowa State University*
Douglas W. Ogle, *Virginia Highlands Community College*
Joel Ostroff, *Brevard Community College*
James Lewis Payne, *Virginia Commonwealth University*
Gary Peterson, *South Dakota State University*
MaryAnn Phillippi, *Southern Illinois University, Carbondale*
R. Douglas Powers, *Boston College*
Robert Raikow, *University of Pittsburgh*
Charles Ralph, *Colorado State University*
Aryan Roest, *California State Polytechnic Univ., San Luis Obispo*
Robert Romans, *Bowling Green State University*
Raymond Rose, *Beaver College*
Richard G. Rose, *West Valley College*
Donald G. Ruch, *Transylvania University*
A. G. Scarbrough, *Towsow State University*
Gail Schiffer, *Kennesaw State University*
John Schmidt, *Ohio State University*
John R. Schrock, *Emporia State University*
Marilyn Shopper, *Johnson County Community College*
John Smarrelli, *Loyola University of Chicago*
Deborah Smith, *Meredith College*
Guy Steucek, *Millersville University*
Ralph Sulerud, *Augsburg College*
Tom Terry, *University of Connecticut*
James Thorp, *Cornell University*
W. M. Thwaites, *San Diego State University*
Michael Torelli, *University of California, Davis*
Michael Treshow, *University of Utah*
Terry Trobec, *Oakton Community College*
Len Troncale, *California State Polytechnic University, Pomona*
Richard Van Norman, *University of Utah*
David Vanicek, *California State University, Sacramento*
Terry F. Werner, *Harris-Stowe State College*
David Whitenberg, *Southwest Texas State University*
P. Kelly Williams, *University of Dayton*
Robert Winget, *Brigham Young University*
Steven Wolf, *University of Missouri, Kansas City*
Harry Womack, *Salisbury State University*
William Yurkiewicz, *Millersville University*

Gil Brum
Larry McKane
Gerry Karp

Brief Table of Contents

Contents

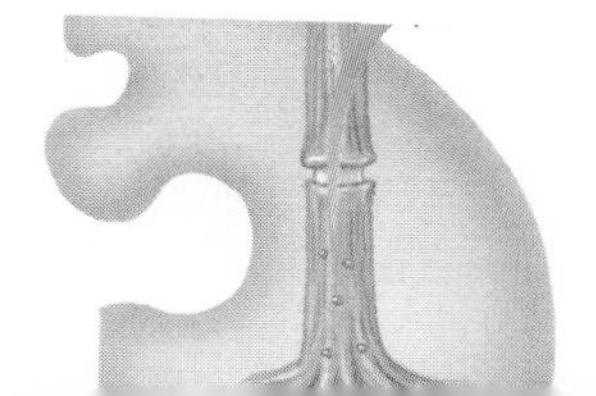

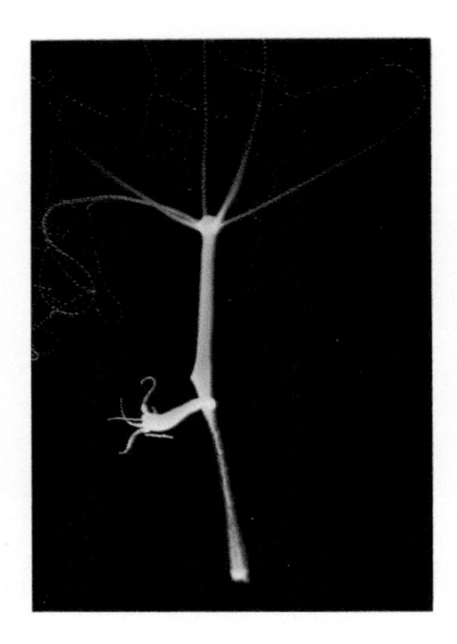

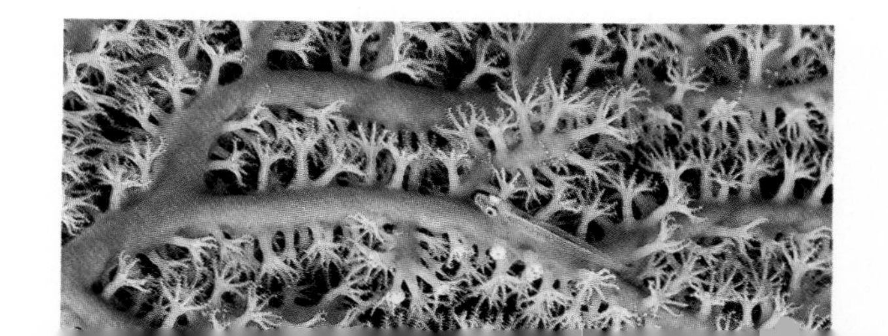

PART 8
Ecology and Animal Behavior 899

To The Student: A User's Guide

Biology is a journey of exploration and discovery, of struggle and breakthrough. It is enlivened by the thrill of understanding not only what living things do but also how they work. We have tried to create such an experience for you.

Excellence in writing, visual images, and broad biological coverage form the core of a modern biology textbook. But as important as these three factors are in making difficult concepts and facts clear and meaningful, none of them reveals the excitement of biology—the adventure that unearths what we know about life. To help relate the true nature of this adventure, we have developed several distinctive features for this book, features that strengthen its biological core, that will engage and hold your attention, that reveal the human side of biology, that enable every reader to understand how science works, that stimulate critical thinking, and that will create the informed citizenship we all hope will make a positive difference in the future of our planet.

Steps to Discovery

The process of science enriches all parts of this book. We believe that students, like biologists, themselves, are intrigued by scientific puzzles. Every chapter is introduced by a "Steps to Discovery" narrative, the story of an investigation that led to a scientific breakthrough in an area of biology which relates to that chapter's topic. The "Steps to Discovery" narratives portray biologists as they really are: human beings, with motivations, misfortunes, and mishaps, much like everyone experiences. We hope these narratives help you better appreciate biological investigation, realizing that it is understandable and within your grasp.

Throughout the narrative of these pieces, the writing is enlivened with scientific work that has provided knowledge and understanding of life. This approach is meant not just to pay tribute to scientific giants and Nobel prize winners, but once again to help you realize that science does not grow by itself. Facts do not magically materialize. They are the products of rational ideas, insight, determination, and, sometimes, a little luck. Each of the "Steps to Discovery" narratives includes a painting that is meant primarily as an aesthetic accompaniment to the adventure described in the essay and to help you form a mental picture of the subject.

STEPS TO DISCOVERY

A Factor Promoting the Growth of Nerves

Rita Levi-Montalcini received her medical degree from the University of Turin in Italy in 1936, the same year that Benito Mussolini began his anti-Semitic campaign. By 1939, as a Jew, Levi-Montalcini had been barred from carrying out research and practicing medicine, yet she continued to do both secretly. As a student, Levi-Montalcini had been fascinated with the structure and function of the nervous system. Unable to return to the university, she set up a simple laboratory in her small bedroom in her family's home. As World War II raged throughout Europe, and the Allies systematically bombed Italy, Levi-Montalcini studied chick embryos in her bedroom, discovering new information about the growth of nerve cells from the spinal cord into the nearby limbs. In her autobiography *In Praise of Imperfection*, she writes: "Every time the alarm sounded, I would carry down to the precarious safety of the cellars the Zeiss binocular microscope and my most precious silver stained embryonic sections." In September 1943, Germa troops arrived in Turin to support the Italian Fascists. Le Montalcini and her family fled southward to Floren where they remained in hiding for the remainder of the w

After the war ended, Levi-Montalcini continued research at the University of Turin. In 1946, she acce an invitation from Viktor Hamburger, a leading expe the development of the chick nervous system, to co Washington University in St. Louis to work with him f semester; she remained at Washington University years.

A chick embryo and one of its nerve cells helped scientists discover nerve growth factor (NGF).

One of Levi-Montalcini's first projects was the reexamination of a previous experiment of Elmer Bueker, a former student of Hamburger's. Bueker had removed a limb from a chick embryo, replaced it with a fragment of a mouse connective tissue tumor, and found that nerve fibers grew into this mass of implanted tumor cells. When Levi-Montalcini repeated the experiment she made an unexpected discovery: One part of the nervous system of these experimental chick embryos—the sympathetic nervous system—had grown five to six times larger than had its counterpart in a normal chick embryo. (The sympathetic nervous system helps control the activity of internal organs, such as the heart and digestive tract.) Close examination revealed that the small piece of tumor tissue that had been grafted onto the embryo had caused sympathetic nerve fibers to grow "wildly" into all of the chick's internal organs, even causing some of the blood vessels to become obstructed by the invasive fibers. Levi-Montalcini hypothesized that the tumor was releasing some soluble substance that induced the remarkable growth of this part of the nervous system. Her hypothesis was soon confirmed by further experiments. She called the active substance **nerve growth factor (NGF).**

The next step was to determine the chemical nature of NGF, a task that was more readily performed by growing the tumor cells in a culture dish rather than an embryo. But Hamburger's laboratory at Washington University did not have the facilities for such work. To continue the project, Levi-Montalcini boarded a plane, with a pair of tumor-bearing mice in the pocket of her overcoat, and flew to Brazil, where she had a friend who operated a tissue culture laboratory. When she placed sympathetic nervous tissue in the proximity of the tumor cells in a culture dish, the nervous tissue sprouted a halo of nerve fibers that grew toward the tumor cells. When the tissue was cultured in the absence of NGF, no such growth occurred.

For the next 2 years, Levi-Montalcini's lab was devoted to characterizing the substance in the tumor cells that possessed the ability to cause nerve outgrowth. The work was carried out primarily by a young biochemist, Stanley Cohen, who had joined the lab. One of the favored approaches to studying the nature of a biological molecule is to determine its sensitivity to enzymes. In order to determine if nerve growth factor was a protein or a nucleic acid, Cohen treated the active material with a small amount of snake venom, which contains a highly active enzyme that degrades nucleic acid. It was then that chance stepped in.

Cohen expected that treatment with the venom ther destroy the activity of the tumor cell fraction was a nucleic acid) or leave it unaffected (if NC protein). To Cohen's surprise, treatment with th *increased* the nerve-growth promoting activity of th rial. In fact, treatment of sympathetic nerve tissue venom alone (in the absence of the tumor extract) i the growth of a halo of nerve fibers! Cohen soon disc why: The snake venom possessed the same nerve factor as did the tumor cells, but at much higher conc tion. Cohen soon demonstrated that NGF was a pr

Levi-Montalcini and Cohen reasoned that since venom was derived from a *modified* salivary gland, other salivary glands might prove to be even better so of the protein. This hypothesis proved to be correct. V Levi-Montalcini and Cohen tested the salivary glands male mice, they discovered the richest source of NGF y source 10,000 times more active than the tumor cells ten times more active than snake venom.

A crucial question remained: Did NGF play a rol the normal development of the embryo, or was its ability stimulate nerve growth just an accidental property of t molecule? To answer this question, Levi-Montalcini a Cohen injected embryos with an antibody against NG which they hoped would inactivate NGF molecules whe ever they were present in the embryonic tissues. The en bryos developed normally, with one major exception: The virtually lacked a sympathetic nervous system. The re searchers concluded that NGF must be important during normal development of the nervous system; otherwise, inactivation of NGF could not have had such a dramatic effect.

By the early 1970s, the amino acid sequence of NGF had been determined, and the protein is now being synthesized by recombinant DNA technology. During the past decade, Fred Gage, of the University of California, has found that NGF is able to revitalize aged or damaged nerve cells in rats. Based on these studies, NGF is currently being tested as a possible treatment of Alzheimer's disease. For their pioneering work, Rita Levi-Montalcini and Stanley Cohen shared the 1987 Nobel Prize in Physiology and Medicine.

A Thematic Approach

Many students are overwhelmed by the diversity of living organisms and the multitude of seemingly unrelated facts that they are forced to learn in an introductory biology course. Most aspects of biology, however, can be thought of as examples of a small number of recurrent themes. Using the thematic approach, the details and principles of biology can be assembled into a body of knowledge that makes sense, and is not just a collection of disconnected facts. Facts become ideas, and details become parts of concepts as you make connections between seemingly unrelated areas of biology, forging a deeper understanding.

All areas of biology are bound together by evolution, the central theme in the study of life. Every organism is the product of evolution, which has generated the diversity of biological features that distinguish organisms from one another and the similarities that all organisms share. From this basic evolutionary theme emerge several other themes that recur throughout the book:

- **Relationship between Form and Function**
- **Biological Order, Regulation, and Homeostatis**
- **Acquiring and Using Energy**
- **Unity Within Diversity**
- **Evolution and Adaptation**

We have highlighted the prevalent recurrence of each theme throughout the text with an icon, shown above. The icons can be used to activate higher thought processes by inviting you to explore how the fact or concept being discussed fits the indicated theme.

Reexamining the Themes

Each chapter concludes with a "Reexamining the Themes" section, which revisits the themes and how they emerge within the context of the chapter's concepts and principles. This section will help you realize that the same themes are evident at all levels of biological organization, whether you are studying the molecular and cellular aspects of biology or the global characteristics of biology.

When two organisms have the same protein, the difference in amino acid sequence of that protein can be correlated with the evolutionary relatedness of the organisms. The amino acid sequence of hemoglobin, for example, is much more similar between humans and monkeys—organisms that are closely related—than between humans and turtles, who are only distantly related. In fact, the evolutionary tree that emerges when comparing the structure of specific proteins from various animals very closely matches that previously constructed from fossil evidence.

The fact that the amino acid sequences of proteins change as organisms diverge from one another reflects an underlying change in their genetic information. Even though a DNA molecule from a mushroom, a redwood tree, and a cow may appear superficially identical, the sequences of nucleotides that make up the various DNA molecules are very different. These differences reflect evolutionary changes resulting from natural selection (Chapter 34).

Virtually all differences among living organisms can be traced to evolutionary changes in the structure of their various macromolecules, originating from changes in the nucleotide sequences of their DNA. (See CTQ #7.)

REEXAMINING THE THEMES

Relationship between Form and Function

The structure of a macromolecule correlates with a particular function. The unbranched, extended nature of the cellulose molecule endows it with resistance to pulling forces, an important property of plant cell walls. The hydrophobic character of lipids underlies many of their biological roles, explaining, for example, how waxes are able to provide plants with a waterproof covering. Protein function is correlated with protein shape. Just as a key is shaped to open a specific lock, a protein is shaped for a particular molecular interaction. For example, the shape of each polypeptide chain of hemoglobin enables a molecule of oxygen to fit perfectly into its binding site. A single alteration in the amino acid sequence of a hemoglobin chain can drastically reduce the molecule's oxygen-carrying capacity.

Biological Order, Regulation, and Homeostasis

Both blood sugar levels and body weight in humans are controlled by complex homeostatic mechanisms. The level of glucose in your blood is regulated by factors acting on the liver, which stimulate either glycogen breakdown (which increases blood sugar) or glycogen formation (which decreases blood sugar). Your body weight is, at least partly, determined by factors emanating from fat cells which either increase metabolic rate (which tends to decrease body weight) or slow down metabolic rate (which tends to increase body weight).

Acquiring and Utilizing Energy

The chemical energy that fuels biological activities is stored primarily in two types of macromolecules: polysaccharides and fats. Polysaccharides, including starch in plants and glycogen in animals, function primarily in the short-term storage of chemical energy. These polysaccharides can be rapidly broken down to sugars, such as glucose, which are readily metabolized to release energy. Gram-for-gram, fats contain even more energy than polysaccharides and function primarily as a long-term storage of chemical energy.

Unity within Diversity

All organisms, from bacteria to humans, are composed of the same four families of macromolecules, illustrating the unity of life—even at the biochemical level. The precise nature of these macromolecules and the ways they are or ganized into higher structures differ from organism to orga nism, thereby building diversity. Plants, for example, po lymerize glucose into starch and cellulose, while anima polymerize glucose into glycogen. Similarly, many protei (such as hemoglobin) are present in a variety of organism but the precise amino acid sequence of the protein va from one species to the next.

Evolution and Adaptation

Evolution becomes very apparent at the mol level when we compare the structure of macromol among diverse organisms. Analysis of the amino a quences of proteins and the nucleotide sequences cleic acids reveals a gradual change over time in th ture of macromolecules. Organisms that are closely have proteins and nucleic acids whose sequences a similar than are those of distantly related organis large degree, the differences observed among di ganisms derives from the evolutionary differenc nucleic acid and protein sequences.

The segregation of alleles and their independent assortment during meiosis increase genotype diversity by promoting new combinations of genes. But the shuffling of existing genes alone does not explain the presence of such a vast diversity of life. If all organisms descended from a common ancestor, with its relatively small complement of genes, where did all the genes present in today's millions of species come from? The answer is mutation.

Most mutant alleles are detrimental; that is, they are more likely to disrupt a well-ordered, smoothly functioning organism than to increase the organism's fitness. For example, a mutation might change a gene so that it produces an inactive enzyme needed for a critical life function. Occasionally, however, one of these stable genetic changes creates an advantageous characteristic t fitness of the offspring. In this way, muta raw material for evolution and the diversi earth.

One of the requirements for genes is stabi remain basically the same from generation t the fitness of organisms would rapidly dete same time, there must be some capacity change; otherwise, there would be no poter tion. Alterations in genes do occur, albeit rar changes (mutations) represent the raw mate tion. (See CTQ #7.)

REEXAMINING THE THEMES

Biological Order, Regulation, and Homeostasis

Mendel discovered that the transmission of genetic factors followed a predictable pattern, indicating that the processes responsible for the formation of gametes, including the segregation of alleles, must occur in a highly ordered manner. This orderly pattern can be traced to the process of meiosis and the precision with which homologous chromosomes are separated during the first meiotic division. Mendel's discovery of independent assortment can also be connected with the first meiotic division, when each pair of homologous chromosomes becomes aligned at the metaphase plate in a manner that is independent of other pairs of homologues.

Unity within Diversity

All eukaryotic, sexually reproducing organisms follow the same "rules" for transmitting inherited traits. Although Mendel chose to work with peas, he could have come to the same conclusions had he studied fruit flies or mice or had he scrutinized a family's medical records on the transmission of certain genetic diseases, such as cystic fibrosis the mechanism by which genes are transmitted is the genes themselves are highly diverse from one to the next. It is this genetic difference among sp forms the very basis of biological diversity

Evolution and Adaptation

Mendel's findings provided a critical link in ou edge of the mechanism of evolution. A key tene theory of evolution is that favorable genetic variat crease the likelihood that an individual will survive productive age and that its offspring will exhibit thes favorable characteristics. Mendel's demonstratio units of inheritance pass from parents to offspring w being blended revealed the means by which advanta traits could be preserved in a species over many ge tions. The subsequent discovery of genetic change by tation revealed how new genes appeared in a popula thus providing the raw material for evolution.

SYNOPSIS

Gregor Mendel discovered the pattern by which inherited traits are transmitted from parents to offspring. Mendel discovered that inherited traits were controlled by pairs of factors (genes). The two factors for a given trait in an individual could be identical (homozygous) or different (heterozygous). In heterozygotes, one of th gene variants (alleles) may be dominant over the othe recessive allele. Because of dominance, the appearanc (phenotype) of the heterozygote (genotype of *Aa*) is identi cal to that of the homozygote with two dominant allele

The Human Perspective

Students will naturally find many ways in which the material presented in any biology course relates to them. But it is not always obvious how you can use biological information for better living or how it might influence your life. Your ability to see yourself in the course boosts interest and heightens the usefulness of the information. This translates into greater retention and understanding.

To accomplish this desirable outcome, the entire book has been constructed with you—the student—in mind. Perhaps the most notable feature of this approach is a series of boxed essays called "The Human Perspective" that directly reveals the human relevance of the biological topic being discussed at that point in the text. You will soon realize that human life, including your own, is an integral part of biology.

ART 2 / *Chemical and Cellular Foundations of Life*

◁ THE HUMAN PERSPECTIVE ▷

Obesity and the Hungry Fat Cell

FIGURE 1 Actor Robert DeNiro in (*left*) a scene from the movie *Raging Bull* and (*right*) a recent photograph.

It has become increasingly clear in recent years that people who are exceedingly overweight—that is, obese—are at increased risk of serious health problems, including heart disease and cancer. By most definitions, a person is obese if he or she is about 20 percent above "normal" or desirable body weight. Approximately 35 percent of adults in the United States are considered obese by this definition, twice as many as at the turn of the century. Among young adults, high blood pressure is five times more prevalent and diabetes three times more prevalent in a group of obese people than in a group of people who are at normal weight. Given these statistics, together with the social stigma facing the obese, there would seem to be strong motivation for maintaining a "normal" body weight. Why, then, are so many of us so overweight? And, why is it so hard to lose unwanted pounds and yet so easy to gain them back? The answers go beyond our fondness for high-calorie foods.

Excess body fat is stored in fat cells (*adipocytes*) located largely beneath the skin. These cells can change their volume more than a hundredfold, depending on the amount of fat they contain. As a person gains body fat, his or her fat cells become larger and larger, accounting for the bulging, sagging body shape. If the person becomes sufficiently overweight, and their fat cells approach their maximum fat-carrying capacity, chemical messages are sent through the blood, causing formation of new fat cells that are "hungry" to begin accumulating their own fat. Once a fat cell is formed, it may expand or contract in volume, but it appears to remain in the body for the rest of the person's life.

Although the subject remains controversial, current research findings suggest that body weight is one of the properties subject to physiologic regulation in humans. Apparently, each person has a particular weight that his or her body's regulatory machinery acts to maintain. This particular value—whether 40 kilograms (80 pounds) or 200 kilograms (400 pounds)—is referred to as the person's **set-point.**

People maintain their body weight at a relatively constant value by balancing energy intake (in the form of food calories) with energy expenditure (in the form of calories burned by metabolic activities or excreted). Obese individuals are thought to have a higher set-point than do persons of normal weight. In many cases, the set-point value appears to have a strong genetic component. For instance, studies reveal there is no correlation between the body mass of adoptees and their adoptive parents, but there is a clear relationship between adoptees and their biological parents, with whom they have not lived.

The existence of a body-weight set-point is most evident when the body weight of a person is "forced" to deviate from the regulated value. Individuals of normal body weight who are fed large amounts of high-calorie foods under experimental conditions tend to gain increasing amounts of weight. If these people cease their energy-rich diets, however, they return quite rapidly to their previous levels, at which point further weight loss stops. This is illustrated by actor Robert DeNiro, who reportedly gained about 50 pounds for the filming of the movie "Raging Bull" (Figure 1), and then lost the weight prior to his next acting role. Conversely, a person who is put on a strict, low-calorie diet will begin to lose weight. The drop in body weight soon triggers a decrease in the person's resting metabolic rate; that is, the amount of calories burned when the person is not engaged in physical activity. The drop in metabolic rate is the body's compensatory measure for the decreased food intake. In other words, it is the body's attempt to halt further weight loss. This effect is particularly pronounced among obese people who diet and lose large amounts of weight: Their pulse rate and blood pressure drop markedly, their fat cells shrink to "ghosts" of their former selves, and they tend to be continually hungry. If these obese individuals go back to eating a *normal* diet, they tend to regain the lost weight rapidly. The drive of these formerly obese persons to increase their food intake is probably a response to chemical signals emanating from the fat cells as they shrink below their previous size.

630 • PART 5 / *Form and Function of Animal Life*

◁ THE HUMAN PERSPECTIVE ▷

Dying for a Cigarette?

e average, smoking cigarettes will cut
imately 6 to 8 years off your life,
han 5 minutes for every cigarette
! Cigarette smoking is the greatest
preventable death in the United
ccording to a 1991 report by the
for Disease Control (CDC),
),000 Americans die each year
ng-related causes. Smoking ac-
87 percent of all lung-cancer
smokers are more susceptible
he esophagus, larynx, mouth,
l bladder than are nonsmok-
reased incidence of lung
mong smokers compared to
hown in Figure 1*a*, and the
l by quitting is shown in
ffects of smoking on lung
Figure 2. Atherosclero-
and peptic ulcers also
greater frequency than
rs. For example, long-
5 times more likely to
terial disease than are
ysema (a condition
ction of lung tissue,
culty in breathing)
mmation of the air-
e prevalent among

ger other people.
sponsible for the
nocent bystand-
re the same air
passive (invol-
own; second-
seriously ill
rs have dou-
y infections
posed to to-
ng married
us; 20 per-
long non-
ttributable to inhaling other

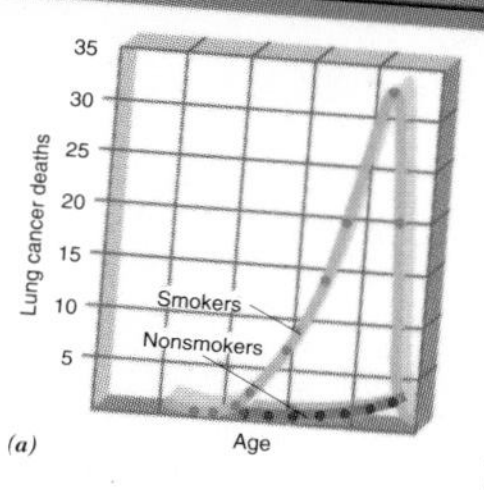

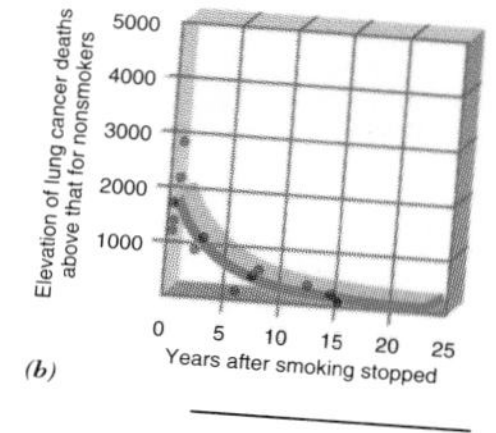

people's tobacco smoke. Another "innocent bystander" is a fetus developing in the uterus of a woman who smokes. Smoking increases the incidence of miscarriage and stillbirth and decreases the birthweight of the infant. Once born, these babies suffer twice as many respiratory infections as do babies of nonsmoking mothers.

Why is smoking so bad for your health? The smoke emitted from a burning cigarette contains more than 2,000 identifiable substances, many of which are either irritants or carcinogens. These compounds include carbon monoxide, sulfur dioxide, formaldehyde, nitrosamines, toluene, ammonia, and radioactive isotopes. Autopsies of respiratory tissues from smokers (and from nonsmokers who have lived for long periods with smokers) show widespread cellular changes, including the presence of precancerous cells (cells that may become malignant, given time) and a marked reduction in the number of cilia that play a vital role in the removal of bacteria and debris from the airways.

Of all the compounds found in tobacco (including smokeless varieties), the most important is nicotine, not because it is carcinogenic, but because it is so addictive. Nicotine is addictive because it acts like a neurotransmitter by binding to certain acetylcholine receptors (page 477), stimulating postsynaptic neurons. The physiological effects of this stimulation include the release of epinephrine, an increase in blood sugar, an elevated heart rate, and the constriction of blood vessels, causing elevated blood pressure. A smoker's nervous system becomes "accustomed" to the presence of nicotine and decreases the output of the natural neurotransmitter. As a result, when a person tries to stop smoking, the sudden absence of nicotine, together with the decreased level of the natural transmitter, decreases stimulation of postsynaptic neurons, which creates a craving for a cigarette—a "nicotine fit." Ex-smokers may be so conditioned to the act of smoking that the craving for cigarettes can continue long after the physiological addiction disappears.

Biolines

The "Biolines" are boxed essays that highlight fascinating facts, applications, and real-life lessons, enlivening the mainstream of biological information. Many are remarkable stories that reveal nature to be as surprising and interesting as any novelist could imagine.

328 • PART 3 / *The Genetic Basis of Life*

◁ BIOLINE ▷

DNA Fingerprints and Criminal Law

On February 5, 1987, a woman and her 2-year-old daughter were found stabbed to death in their apartment in the New York City borough of the Bronx. Following a tip, the police questioned a resident of a neighboring building. A small bloodstain was found on the suspect's watch, which was sent to a laboratory for DNA fingerprint analysis. The DNA from the white blood cells in the stain was amplified using the PCR technique and was digested with a restriction enzyme. The restriction fragments were then separated by electrophoresis, and a pattern of labeled fragments was identified with a radioactive probe. The banding pattern produced by the DNA from the suspect's watch was found to be a perfect match to the pattern produced by DNA taken from one of the victims. The results were provided to the opposing attorneys, and a pretrial hearing was called in 1989 to discuss the validity of the DNA evidence.

During the hearing, a number of expert witnesses for the prosecution explained the basis of the DNA analysis. According to these experts, no two individ-
als, with the exception of identical twins,
ave the same nucleotide sequence in their
NA. Moreover, differences in DNA se-
ence can be detected by comparing the
gths of the fragments produced by re-
tion-enzyme digestion of different
A samples. The patterns produce a
A fingerprint" (Figure 1) that is as
e to an individual as is a set of con-
nal fingerprints lifted from a glass. In
NA fingerprints had already been
more than 200 criminal cases in the
States and had been hailed as the
portant development in forensic
(the application of medical facts to legal problems) in decades. The widespread use of DNA fingerprinting evidence in court had been based on its general acceptability in the scientific community. According to a report from the company performing the DNA analysis, the likelihood that the same banding patterns could be obtained by chance from two *different* individuals in the community was only one in 100 million.

FIGURE 1

Alec Jeffreys of the University of Leicester, England, examining a DNA fingerprint. Jeffreys was primarily responsible for developing the DNA fingerprint technique and was the scientist who confirmed the death of Josef Mengele.

What made this case (known as the Castro case, after the defendant) memorable and distinct from its predecessors was that the defense also called on expert witnesses to scrutinize the data and to present their opinions. While these experts
firmed the capability of DNA finger
ing to identify an individual out of a
population, they found serious tech
flaws in the analysis of the DNA sam
used by the prosecution. In an unpre
dented occurrence, the experts who
earlier testified *for the prosecution* agr
that the DNA analysis in this case
unreliable and should not be used as
dence! The problem was not with the te
nique itself but in the way it had b
carried out in this particular case. Con
quently, the judge threw out the eviden

In the wake of the Castro case, the u
of DNA fingerprinting to decide guilt
innocence has been seriously questione
Several panels and agencies are working
formulate guidelines for the licensing
forensic DNA laboratories and the certif
cation of their employees. In 1992, a pane
of the National Academy of Sciences re
leased a report endorsing the general reli
ability of the technique but called for the
institution of strict standards *to be set by
scientists.*

Meanwhile, another issue regarding DNA fingerprinting has been raised and hotly debated. Two geneticists, Richard Lewontin of Harvard University and Daniel Hartl of Washington University, coauthored a paper published in December 1991, suggesting that scientists do not have enough data on genetic variation within different racial or ethnic groups to calculate the odds that two individuals—a suspect and a perpetrator of the crime—are one and the same on the basis of an identical DNA fingerprint. The matter remains an issue of great concern in both the scientific and legal communities and has yet to be resolved.

254 • PART 3 / *The Genetic Basis of Life*

◁ BIOLINE ▷

The Fish That Changes Sex

In vertebrates, gender is generally a biologically inflexible commitment: An individual develops into either a male or a female as dictated by the sex chromosomes acquired from one's parents. Yet, even among vertebrates, there are organisms that can reverse their sexual commitment. The Australian cleaner fish (Figure 1), a small animal that sets up "cleaning stations" to which larger fishes come for parasite removal, can change its gender in response to environmental demands. Most male cleaner fish travel alone rather than with a school. Except for a single male, schools of cleaner fish are comprised entirely of females. Although it might seem logical to conclude that maleness engenders solo travel, it is actually the other way around: Being alone fosters maleness. A cleaner fish that develops away from a school *becomes* a male, whereas the same fish developing in a school would have become a female.

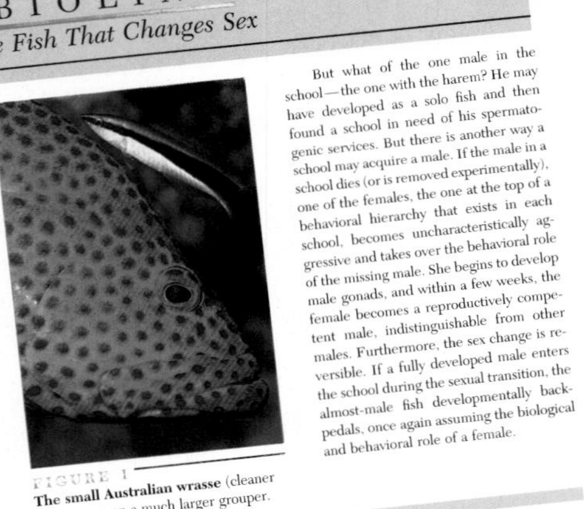

FIGURE 1

The small Australian wrasse (cleaner fish) is seen on a much larger grouper.

But what of the one male in the school—the one with the harem? He may have developed as a solo fish and then found a school in need of his spermatogenic services. But there is another way a school may acquire a male. If the male in a school dies (or is removed experimentally), one of the females, the one at the top of a behavioral hierarchy that exists in each school, becomes uncharacteristically aggressive and takes over the behavioral role of the missing male. She begins to develop male gonads, and within a few weeks, the female becomes a reproductively competent male, indistinguishable from other males. Furthermore, the sex change is reversible. If a fully developed male enters the school during the sexual transition, the almost-male fish developmentally backpedals, once again assuming the biological and behavioral role of a female.

Not all organisms follow the mammalian pattern of sex determination. In some animals, most notably birds, the opposite pattern is found: The female's cells have an X and a Y chromosome, while the male's cells have two Xs. An exception to this rule of a strict relation between sex and chromosomes is discussed in the Bioline: The Fish That Changes Sex. Although some plants possess sex chromosomes and gender distinctions between individuals, most have only autosomes; consequently, each individual produces both male and female parts.

SEX LINKAGE

For fruit flies and humans alike, there are hundreds of genes on the X chromosome that have no counterpart on the smaller Y chromosome. Most of these genes have nothing to do with determining gender, but their effect on phenotype usually *depends on* gender. For example, in females, a recessive allele on one X chromosome will be masked (and not expressed) if a dominant counterpart resides on the other X chromosome. In males, it only takes one recessive allele on the single X chromosome to determine the individual's phenotype since there is no corresponding allele on the Y chromosome. Inherited characteristics determined by genes that reside on the X chromosome are called **X-linked characteristics.**

So far, some 200 human X-linked characteristics have been described, many of which produce disorders that are found almost exclusively in men. These include a type of heart-valve defect (*mitral stenosis*), a particular form of mental retardation, several optical and hearing impairments, muscular dystrophy, and red-green colorblindness (Figure 13-8).

One X-linked recessive disorder has altered the course of history. The disease is **hemophilia,** or "bleeder's disease," a genetic disorder characterized by the inability to produce a clotting factor needed to halt blood flow quickly following an injury. Nearly all hemophiliacs are males. Although females can inherit two recessive alleles for hemophilia, this occurrence is extremely rare. In general, women who have acquired the rare defective allele are heterozygous **carriers** for the disease. The phenotype of a carrier

Bioethics Essays

Several ethical issues are discussed in the Bioethics essays which add provocative pauses throughout the text. Biological Science does not operate in a vacuum but has profound consequences on the general community. Because biologists study life, the science is peppered with ethical considerations. The moral issues discussed in these essays are neither simple nor easy to resolve, and we do not claim to have any certain answers. Our goal is to encourage you to consider the bioethical issues that you will face now and in the future.

Coordinating the Organism: The Role of the Nervous System / CHAPTER 23 • 489

BIOETHICS

Blurring the Line between Life and Death

By **ARTHUR CAPLAN**
Division of the Center for Biomedical Ethics at the University of Minnesota

Theresa Ann Campo Pearson didn't have a very long life. When she died in 1992, she was only 10 days old. Despite her short life, she became the center of a very strange, sad, and wrenching ethical controversy. Theresa died because her brain had failed to form. She had anencephaly, a condition in which only the brainstem, located at the top of the spinal cord, is present. Her parents wanted to donate Theresa's organs; the courts said no. Some people found it strange that Theresa's parents, Laura Campo and Justin Pearson, did not get their way. Why not allow donation, when every day in North America a baby dies because there is no heart, lung, or liver available for transplantation?

Anencephaly is best described as completely "unabling," not disabling. Children born with anencephaly cannot think, feel, sense, or be aware of the world. Many are stillborn; the majority of the rest die within days of birth. A mere handful live for a few weeks. Theresa's parents knew all this. But rather than abort the pregnancy, they chose to have their baby. In fact, the baby was born by Caesarean section, at least partly in the hope that it would be born alive, thereby making organ donation possible. When Theresa died at Broward General Medical Center in Fort Lauderdale, Florida, however, no organs were taken. Two Florida courts ruled that the baby could not be used as a source of organs unless she was brain-dead, and Theresa Ann Campo was never pronounced brain-dead.

Brain death refers to a situation in which the brain has irreversibly lost all function and activity. Babies born with anencephaly have some brain function in their brainstem so, while they cannot think or feel, they are alive. According to Florida law—and the law in more than 40 other states—only those individuals declared brain-dead can donate organs. The courts of Florida had no other option but to deny the request for organ donation.

One obvious solution is to change the law so that states could decide that organs can be removed upon parental consent from either those who are born brain-dead or from babies who are born with anencephaly. Another solution is to rewrite the definition of death to say that death occurs either when the brain has totally ceased to function or if a baby is born anencephalic. Do you feel that either of these changes should be made? Some may argue that medicine will fudge the line between life and death in order to get organs for transplant. Do you agree with this concern? How do you think redefining death will affect a person's decision to check off the donation box on the back of a driver's license? Do you think people may worry that if they are known to be potential donors they won't be aggressively treated at the hospital? In your opinion, would changing the definition of death to include anencephaly be beneficial or deleterious?

Like the brain, the spinal cord is composed of white matter (myelinated axons) and gray matter (dendrites and cell bodies). However, the arrangement of these types of matter is reversed in the spinal cord, compared to their arrangement in the brain: The spinal cord's white matter surrounds the gray matter (Figure 23-16).

The human central nervous system is the most complex and highly evolved assembly of matter. Among its functions are the processing of sensory information collected from both the external and internal environment; the regulation of internal physiological activities; the coordination of complex motor activities; and the endowment of such intangible "mental" qualities as emotions, creativity, language, and the ability to think, learn, and remember. (See CTQ #6.)

ARCHITECTURE OF THE PERIPHERAL NERVOUS SYSTEM

The peripheral nervous system provides the neurological bridge between the central nervous system and the various parts of the body. The peripheral nervous system is made up of paired nerves that extend into the periphery from the CNS at various levels along the body. Each nerve is composed of a large bundle of myelinated axons surrounded by a connective tissue sheath. Twelve pairs of **cranial nerves** emerge from the central stalk of the human brain, and 31 pairs of **spinal nerves** extend from the spinal cord out between the vertebrae of humans (Figure 23-16). For the most part, the cranial nerves *innervate* (supply nerves to) tissues and organs of the head and neck, whereas the spinal nerves innervate the chest, abdomen, and limbs.

Additional Pedagogical Features

We have worked to assure that each chapter in this book is an effective teaching and learning instrument. In addition to the pedagogical features discussed above, we have included some additional tried-and-proven-effective tools.

KEY POINTS

Key points follow each major section and offer a condensation of the relevant facts and details as well as the concepts discussed. You can use these key points to reaffirm your understanding of the previous reading or to alert you to misunderstood material before moving on to the next topic. Each key point is tied to a Critical Thinking Question found at the end of the chapter; together, they encourage you to analyze the information, taking it beyond mere memorization.

Plant Tissues and Organs / CHAPTER 18 • 361

Many plants replenish old and dying cells with vigorous new cells. But since each plant cell has a surrounding cell wall (Chapter 7) old plant cells do not just wither and disappear when they die. Instead, dead plant cells leave cellular "skeletons" where they once lived. As a result, the longer a plant lives, the more complex its anatomy becomes. **Annuals** are plants that live for 1 year or less, such as corn and marigolds. Because they live for such a brief period, these plants do not completely replace old cells. As a result, annuals are anatomically less complex than are **biennials**—plants that live for 2 years—and **perennials**—herbs, shrubs, and trees that live longer than 2 years. Biennials (carrots, Queen Anne's lace) and perennials (rosebushes, apple trees) are able to live longer than annuals because they produce new cells to replace those that cease functioning or die, providing a continual supply of young, vigorous cells.

In this chapter, we will focus on the body construction of flowering plants, the most familiar, most evolutionarily advanced, and structurally complex of any group in the plant kingdom. All flowering plants are **vascular plants;** that is, they contain specialized cells that circulate water, minerals, and food (organic molecules) throughout the plant. Botanists divide flowering plants into two main groups: **dicotyledons,** or dicots (*di* = two, *cotyledon* = embryonic seed leaf), and **monocotyledons,** or monocots (*mono* = one). Table 18-1 illustrates the many differences that distinguish dicots from monocots and will be used as a reference throughout the chapter.

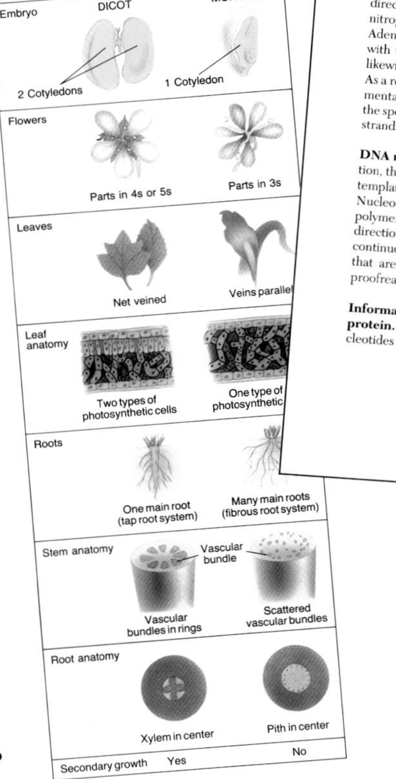

TABLE 18-1

Dicot and Monocot Comparison	DICOT	MONOCOT
Embryo	2 Cotyledons	1 Cotyledon
Flowers	Parts in 4s or 5s	Parts in 3s
Leaves	Net veined	Veins paralle
Leaf anatomy	Two types of photosynthetic cells	One type of photosynthetic
Roots	One main root (tap root system)	Many main roots (fibrous root system)
Stem anatomy	Vascular bundles in rings	Scattered vascular bundles
Root anatomy	Xylem in center	Pith in center
Secondary growth	Yes	No

SHOOTS AND ROOTS

The flowering plant body is a study in contradictions. A typical plant grows through the soil and the air simultaneously, two very different habitats with very different conditions. As a result, the two main parts of the plant differ dramatically in form (anatomy) and function (physiology): The underground **root system** anchors the plant in the soil and absorbs water and nutrients, while the aerial **shoot system** absorbs sunlight and gathers carbon dioxide for photosynthesis (Figure 18-2). The shoot system also produces stems, leaves, flowers, and fruits. Interconnected vascular tissues transport materials between the aerial shoot system and the underground root system. These connections allow water and minerals absorbed by the root to be conducted to shoot tissues, and for food produced by the shoot to be transported to root tissues. We will discuss the various components of these two systems in more detail later in the chapter.

Over 90 percent of all plant species are flowering plants. Flowering plants are the most recently evolved plant group, having undergone rapid evolution during the past 1 million to 2 million years as environmental conditions on land became more variable. (See CTQ #2.)

SYNOPSIS

The synopsis section offers a convenient summary of the chapter material in a readable narrative form. The material is summarized in concise paragraphs that detail the main points of the material, offering a useful review tool to help reinforce recall and understanding of the chapter's information.

288 • PART 3 / The Genetic Basis of Life

the corresponding polypeptide. The cumulative effect of gradual changes in polypeptides over evolutionary time has been the generation of life's diversity.

Evolution and Adaptation

Evolutionary change from generation to generation depends on genetic variability. Much of this variability arises from reshuffling maternal and paternal genes during meiosis, but somewhere along the way *new* genetic information must be introduced into the population. Ne netic information arises from mutations in existing Some of these mutations arise during replication; occur as the result of unrepaired damage as the DNA "sitting" in a cell. Mutations that occur in an indivi germ cells can be considered the raw material on natural selection operates; whereas harmful mutation duce offspring with a reduced fitness, beneficial muta produce offspring with an increased fitness.

SYNOPSIS

Experiments in the 1940s and 1950s established conclusively that DNA is the genetic material. These experiments included the demonstration that DNA was capable of transforming bacteria from one genetic strain to another; that bacteriophages injected their DNA into a host cell during infection; and that the injected DNA was transmitted to the bacteriophage progeny.

DNA is a double helix. DNA is a helical molecule consisting of two chains of nucleotides running in opposite directions, with their backbones on the outside, and the nitrogenous bases facing inward like rungs on a ladder. Adenine-containing nucleotides on one strand always pair with thymine-containing nucleotides on the other strand, likewise for guanine- and cytosine-containing nucleotides. As a result, the two strands of a DNA molecule are complementary to one another. Genetic information is encoded in the specific linear sequence of nucleotides that make up the strands.

DNA replication is semiconservative. During replication, the double helix separates, and each strand serves as a template for the formation of a new, complementary strand. Nucleotide assembly is carried out by the enzyme DNA polymerase, which moves along the two strands in opposite directions. As a result, one of the strands is synthesized continuously, while the other is synthesized in segments that are covalently joined. Accuracy is maintained by a proofreading mechanism present within the polymerase.

Information flows in a cell from DNA to RNA to protein. Each gene consists of a linear sequence of nucleotides that determines the linear sequence of amino acids in a polypeptide. This is accomplished in two ma steps: transcription and translation.

During transcription, the information spelled out the gene's nucleotide sequence is encoded in a mo cule of messenger RNA (mRNA). The mRNA contai a series of codons. Each codon consists of three nucleotid Of the 64 possible codons, 61 specify an amino acid, and t other 3 stop the process of protein synthesis.

During translation, the sequence of codons in th mRNA is used as the basis for the assembly of a chai of specific amino acids. Translating mRNA message occurs on ribosomes and requires tRNAs, which serve a decoders. Each tRNA is folded into a cloverleaf structur with an anticodon at one end—which binds to a complementary codon in the mRNA—and a specific amino acid a the other end—which becomes incorporated into the growing polypeptide chain. Amino acids are added to their appropriate tRNAs by a set of enzymes. The sequential interaction of charged tRNAs with the mRNA results in the assembly of a chain of amino acids in the precise order dictated by the DNA.

Mutation is a change in the genetic message. Gene mutations may occur as a single nucleotide substitution, which leads to the insertion of an amino acid different from that originally encoded. In contrast, the addition of one or two nucleotides throws off the reading frame of the ribosome as it moves along the mRNA, leading to the incorporation of incorrect amino acids "downstream" from the point of mutation. Exposure to mutagens increases the rate of mutation.

REVIEW QUESTIONS

Along with the synopsis, the Review Questions provide a convenient study tool for testing your knowledge of the facts and processes presented in the chapter.

STIMULATING CRITICAL THINKING

Each chapter contains as part of its end material a diverse mix of Critical Thinking Questions. These questions ask you to apply your knowledge and understanding of the facts and concepts to hypothetical situations in order to solve problems, form hypotheses, and hammer out alternative points of view. Such exercises provide you with more effective thinking skills for competing and living in today's complex world.

224 • PART 2 / *Chemical and Cellular Foundations of Life*

Key Terms

zygote (p. 214)
meiosis (p. 214)
life cycle (p. 214)
germ cell (p. 214)
somatic cell (p. 214)
meiosis I (p. 216)
reduction division (p. 216)
synapsis (p. 216)
tetrad (p. 216)
crossing over (p. 216)
genetic recombination (p. 216)
synaptonemal complex (p. 218)
maternal chromosome (p. 219)
paternal chromosome (p. 219)
independent assortment (p. 219)
meiosis II (p. 219)

Review Questions

1. Match the activity with the phase of meiosis in which it occurs.

   a. synapsis
   b. crossing over
   c. kinetochores split
   d. independent assortment
   e. homologous chromosomes separate
   f. cytokinesis

   1. prophase I
   2. metaphase I
   3. anaphase I
   4. telophase I
   5. prophase II
   6. anaphase II
   7. telophase II

2. How do crossing over and independent assortment increase the genetic variability of a species?
3. Why is meiosis I (and not meiosis II) referred to as the reduction division?
4. Suppose that one human sperm contains x amount of DNA. How much DNA would a cell just entering meiosis contain? A cell entering meiosis II? A cell just completing meiosis II? Which of these three cells would have a haploid number of chromosomes? A diploid number of chromosomes?

Critical Thinking Questions

1. Why are disorders, such as Down syndrome, that arise from abnormal chromosome numbers, characterized by a number of seemingly unrelated abnormalities?
2. A gardener's favorite plant had white flowers and long seed pods. To add some variety to her garden, she transplants some plants of the same type, but with pink flowers and short seed pods from her neighbor's garden. To her surprise, in a few generations, she grows plants with white flowers and short seed pods and plants with pink flowers and long seed pods, as well as the original combinations. What are two ways in which these new combinations could have arisen?
3. Set up the meiosis template in the diagram below on a large sheet of paper. Then use pieces of colored yarn or pipe cleaners to simulate chromosomes and make a model of the phases of meiosis. *(See template on opposite page)*
4. Would you expect two genes on the same chromosome, such as yellow flowers and short stems, always to be exchanged during crossing over? How might they remain together *in spite of* crossing over?
5. Suppose paternal chromosomes always lined up on the same side of the metaphase plate of cells in meiosis I. How would this affect genetic variability of offspring? Would they all be identical? Why or why not?

Additional Readings

Chandley, A. C. 1988. Meiosis in man. *Trends in Gen.* 4:79–83. (Intermediate)

Hsu, T. C. 1979. *Human and mammalian cytogenetics.* New York: Springer-Verlag. (Intermediate)

John, B. 1990. *Meiosis.* New York: Cambridge University Press. (Advanced)

Moens, P. B. 1987. *Meiosis.* Orlando: Academic. (Advanced)

Patterson, D. 1987. The causes of Down syndrome. *Sci. Amer.* Feb:52–60. (Intermediate-Advanced)

White, M. J. D. 1973. *The chromosomes.* Halsted. (Advanced)

ADDITIONAL READINGS

Supplementary readings relevant to the Chapter's topics are provided at the end of every chapter. These readings are ranked by level of difficulty (introductory, intermediate, or advanced) so that you can tailor your supplemental readings to your level of interest and experience.

Careers in Biology

*T*he appendices of this edition include "Careers in Biology," a frequently overlooked aspect of our discipline. Although many of you may be taking biology as a requirement for another major (or may have yet to declare a major), some of you are already biology majors and may become interested enough to investigate the career opportunities in life sciences. This appendix helps students discover how an interest in biology can grow into a livelihood. It also helps the instructor advise students who are considering biology as a life endeavor.

APPENDIX

D

Careers in Biology

Although many of you are enrolled in biology as a requirement for another major, some of you will become interested enough to investigate the career opportunities in life sciences. This interest in biology can grow into a satisfying livelihood. Here are some facts to consider:

- Biology is a field that offers a very wide range of possible science careers
- Biology offers high job security since many aspects of it deal with the most vital human needs: health and food
- Each year in the United States, nearly 40,000 people obtain bachelor's degrees in biology. But the number of newly created and vacated positions for biologists is increasing at a rate that exceeds the number of new graduates. Many of these jobs will be in the newer areas of biotechnology and bioservices.

Biologists not only enjoy job satisfaction, their work often changes the future for the better. Careers in medical biology help combat diseases and promote health. Biologists have been instrumental in preserving the earth's life-supporting capacity. Biotechnologists are engineering organisms that promise dramatic breakthroughs in medicine, food production, pest management, and environmental protection. Even the economic vitality of modern society will be increasingly linked to biology.

Biology also combines well with other fields of expertise. There is an increasing demand for people with backgrounds or majors in biology complexed with such areas as business, art, law, or engineering. Such a distinct blend of expertise gives a person a special advantage.

The average starting salary for all biologists with a Bachelor's degree is $22,000. A recent survey of California State University graduates in biology revealed that most were earning salaries between $20,000 and $50,000. But as important as salary is, most biologists stress job satisfaction, job security, work with sophisticated tools and scientific equipment, travel opportunities (either to the field or to scientific conferences), and opportunities to be creative in their job as the reasons they are happy in their career.

Here is a list of just a few of the careers for people with degrees in biology. For more resources, such as lists of current openings, career guides, and job banks, write to Biology Career Information, John Wiley and Sons, 605 Third Avenue, New York, NY 10158.

A SAMPLER OF JOBS THAT GRADUATES HAVE SECURED IN THE FIELD OF BIOLOGY*

Agricultural Biologist
Agricultural Economist
Agricultural Extension Officer
Agronomist
Amino-acid Analyst
Analytical Biochemist
Anatomist
Animal Behavior Specialist
Anticancer Drug Research Technician
Antiviral Therapist
Arid Soils Technician
Audio-neurobiologist
Author, Magazines & Books
Behavioral Biologist
Bioanalyst
Bioanalytical Chemist
Biochemical/Endocrine Toxicologist
Biochemical Engineer
Pharmacology Distributor
Pharmacology Technician
Biochemist
Biogeochemist
Biogeographer
Biological Engineer
Biologist
Biomedical Communication Biologist
Biometerologist
Biophysicist
Biotechnologist
Blood Analyst
Botanist
Brain Function Researcher
Cancer Biologist
Cardiovascular Biologist
Cardiovascular/Computer Specialist
Chemical Ecologist
Chromatographer
Clinical Pharmacologist
Coagulation Biochemist
Cognitive Neuroscientist
Computer Scientist
Dental Assistant
Ecological Biochemist
Electrophysiology/Cardiovascular Technician
Energy Regulation Officer
Environmental Biochemist
Environmental Center Director
Environmental Engineer
Environmental Geographer
Environmental Law Specialist
Farmer
Fetal Physiologist
Flavorist
Food Processing Technologist
Food Production Manager
Food Quality Control Inspector
Flower Grower
Forest Ecologist
Forest Economist
Forest Engineer
Forest Geneticist
Forest Manager

Study Guide

Written by Gary Wisehart and Michael Leboffe of San Diego City College, the *Study Guide* has been designed with innovative pedagogical features to maximize your understanding and retention of the facts and concepts presented in the text. Each chapter in the *Study Guide* contains the following elements.

Concepts Maps

In Chapter 1 of the *Study Guide,* the beginning of a concept map stating the five themes is introduced. In each subsequent chapter, the concept map is expanded to incorporate topics covered in each chapter as well as the interconnections between chapters and the five themes. "Connector" phrases are used to link the concepts and themes, and the text icons representing the themes are incorporated into the concept maps.

Go Figure!

In each chapter, questions are posed regarding the figures in the text. Students can explore their understanding of the figures and are asked to think critically about the figures based on their understanding of the surrounding text and their own experiences.

Self-Tests

Each chapter includes a set of matching and multiple-choice questions. Answers to the Study Guide questions are provided.

Concept Map Construction

The student is asked to create concept maps for a group of terms, using appropriate connector phrases and adding terms as necessary.

Laboratory Manual

Biology: Exploring Life, Second Edition is supplemented by a comprehensive *Laboratory Manual* containing approximately 60 lab exercises chosen by the text authors from the National Association of Biology Teachers. These labs have been thoroughly class-tested and have been assembled from various scientific publications. They include such topics as

- Chaparral and Fire Ecology: Role of Fire in Seed Germination *(The American Biology Teacher)*
- A Model for Teaching Mitosis and Meiosis *(American Biology Teacher)*
- Laboratory Study of Climbing Behavior in the Salt Marsh Snail *(Oceanography for Landlocked Classrooms)*
- Down and Dirty DNA Extraction *(A Sourcebook of Biotechnology Activities)*
- Bioethics: The Ice-Minus Case *(A Sourcebook of Biotechnology Activities)*
- Using Dandelion Flower Stalks for Gravitropic Studies *(The American Biology Teacher)*
- pH and Rate of Enzymatic Reactions *(The American Biology Teacher)*

CHAPTER 33

Mechanisms of Evolution

STEPS TO DISCOVERY
Silent Spring Revisited

STEPS TO DISCOVERY

Silent Spring Revisited

In 1939, Paul Muller, a researcher at a Swiss pharmaceutical company, discovered that the compound dichloro-diphenyl-trichloro-ethane (DDT) was a very effective insecticide. When the United States entered World War II, two insect-borne diseases presented a serious threat to troops: typhus fever, which was common in areas of Europe and was spread by lice; and malaria, which was common in the Pacific and was spread by mosquitos. When an epidemic of typhus threatened to break out in Naples, Italy, in 1943, a powdered preparation of DDT was sprayed under the clothing of over 3 million troops and civilians. For the first time in human history, the spread of this deadly disease was arrested; DDT was hailed as a "miracle" compound. The insecticide proved equally effective against mosquitoes in the Pacific, where it was sprayed from the air over entire islands.

After the war, DDT became available in large quantities to civilians. In 1957, planes flying over marshy areas of Massachusetts, spraying DDT in order to kill mosquitoes, happened to spray the property of a resident who kept a 2-acre bird sanctuary. In addition to killing the insects in the area, the insecticide killed the birds, leaving the landscape ghostly silent. The resident wrote a letter to a friend, Rachel Carson, a biologist and author of several widely acclaimed

Although spraying DDT killed most mosquitoes, the natural genetic variation in the population allowed a small percentage of

books on the sea and its inhabitants. Carson decided to look into the matter. The deeper she became immersed in the scientific literature on the effects of DDT and related pesticides, the more convinced she became that she had to warn the public of their dangers.

After 4 years of research, Carson wrote *Silent Spring*, a book that documented the devastating effects pesticides were having on the wildlife of the world and the problems they were creating for future insect eradication programs. The book inspired President Kennedy to establish a commission to regulate the use of pesticides. Congress began holding hearings on the subject, and environmental concern groups were established. These events culminated in the establishment of the Environmental Protection Agency in 1970, which banned the use of DDT in 1972.

In addition to changing environmental politics, *Silent Spring* provided documented evidence of the role of natural selection in shaping the characteristics of animal populations. Carson described how insects were changing over time to become resistant to pesticides, particularly DDT. She wrote: "Darwin himself could scarcely have found a better example of the operation of natural selection than is provided by the way the mechanism of [pesticide] resistance operates."

A pesticide is a powerful selective agent. In any given population of insects, some individuals have a combination of genes that makes them less susceptible to harmful chemicals than do other individuals. Those individuals that are susceptible to the pesticide die off, removing the genes that confer susceptibility from the population and leaving resistant individuals to repopulate the species' ranks. Initially, only a small fraction of the insect population had the specific combination of genes that made them resistant to potent pesticides. Consequently, when DDT was first used, most of the insects died off. Since insects can produce tremendous numbers of offspring in very short periods, however, it was only a matter of a few years before highly resistant individuals dominated the population.

Moreover, favorable traits, such as pesticide resistance, need not stop at a population's boundaries. A 1991 study investigating pesticide resistance in mosquitoes of the species *Culex pipiens* found that individual insects from around the world carry precisely the same resistance-promoting genetic alteration. This finding strongly suggests that resistance in different populations is not due to independent mutations within each population; rather, resistance can spread rapidly from population to population when individuals carrying the beneficial genes migrate to new environments.

Species that lack the ability to cope with environmental changes will shrink in number or even become extinct, an event chronicled in Carson's book. Carson noted that, while some insect populations were evolving pesticide resistance, many bird populations were being decimated. Studies showed the birds were eating pesticide-contaminated prey (including insects, earthworms, and fish), and the toxins were building to high concentrations in the birds' fatty tissues. The accumulating DDT prevented many birds from producing healthy offspring: Birds that ingested DDT laid fewer eggs; the eggs had thinner shells and were often broken in the nest; and the chicks that hatched were so loaded with pesticide residues that they often failed to survive. Unlike insects, none of the members of the bird populations possessed genes that conferred resistance to pesticides; therefore, there were no resistant individuals for natural selection to favor. Many people became concerned that some of these bird species, including the American eagle and the peregrine falcon, were being pushed to extinction. Fortunately, since 1970, these bird species have actually grown in number, largely as a result of a ban on the use of DDT.

Concern over the effects of DDT is not limited to birds. Even though DDT has been banned in many countries, the chemical residues of the pesticide can remain stored for decades in the fatty tissues of the human body. A report in 1993 indicated that women with high levels of DDT in their body had four times the risk of contracting breast cancer compared to women with the lowest levels of the pesticide. The earlier use of DDT may be one of the reasons for the puzzling rise in breast cancer rates in the past few decades.

the mosquitoes to survive and repopulate the species.

"A fish out of water" can't survive very long, or so we might think. Yet the mudskipper is a fish that not only survives long treks across mud flats, it even climbs trees (Figure 33-1). In water, the mudskipper's fins and gills work just like those of a typical fish: Its fins propel and steer the mudskipper through the water, and its gills extract dissolved oxygen. But how does the mudskipper remain alive on land with a body and breathing machinery that are so unmistakably adapted for life in the water?

Unlike other fishes, the mudskipper's gills and fins have modifications that enable it to survive and move on land. For its respiration, the mudskipper packs a supply of water into its bulging gill pouches, from which it extracts oxygen; the pouches act like a scuba tank in reverse. For its motility, the mudskipper's reinforced forefins serve as stubby arms for crawling across the mud or shimmying up tree trunks in search of snails for a meal.

A hobbling mudskipper with water-engorged pouches illustrates how structures originally adapted for one way of life can become refashioned for new functions. The mudskipper's makeshift legs and water bags are modifications of structures that were originally adapted for aquatic life. The fact that the mudskipper possesses these structures is evidence that its ancestors lived strictly underwater and that those ancestors were something other than mudskippers. Every species has a history of ancestors that possessed features and behaviors different from those the existing species possess. To trace the history of change in an organism's ancestors is to follow its course of evolution.

FIGURE 33-1
The mudskipper is a fish that can live for hours out of water. Strengthened forefins and modified gill chambers are evidence that the mudskipper evolved from ancestors that were strictly aquatic.

THE BASIS OF EVOLUTIONARY CHANGE

How did there come to be so many kinds of organisms, each with unique characteristics that enable it to survive in one or more of the earth's many diverse habitats? How can we explain the fact that there were organisms living in the past that possessed characteristics very different from those of organisms living today? And why were the earth's first organisms much simpler than are many of today's organisms? **Evolution**—the theory that a population of organisms changes as the generations pass—provides the answers to these questions. Evolution explains how modern animals and plants evolved from more primitive ancestors, which, in turn, evolved from even more primitive types, and so on, back to the first appearance of life, billions of years ago. The fact that all the diverse organisms present on earth today arose from a common ancestor explains why they have the same basic mechanism for the storage and utilization of genetic information, many of the same types of cellular organelles, and similar types of enzymes and metabolic pathways. These shared characteristics were present in the earliest organisms and were retained among their descendants.

Contrary to popular belief, the concept of evolution did not originate with Darwin. In the eighteenth century, a few naturalists had considered the *possibility* of descent with modification, but no one had proposed a convincing explanation as to *how* evolution could occur. Furthermore, strong religious opposition to the concept of evolution contributed to a general lack of interest. Consequently, the theory of evolution was not widely accepted until the middle of the nineteenth century, at which time Charles Darwin suggested a plausible mechanism—natural selection—and collected substantial evidence to support the contention that evolution had indeed occurred (page 21). In the twentieth century, as the principles of genetics became better understood, biologists discovered mechanisms other than natural selection that cause organisms to change from generation to generation. The synthesis of Darwin's original ideas on evolution by natural selection and those of modern genetic theory is referred to as either *neodarwinism* or the *modern synthesis.*

THE GENETIC COMPOSITION OF POPULATIONS

Individual organisms are born, mature, and eventually die. Along the way, an individual may change, but it does not evolve. Rather, it is the **species**—a group of interbreeding organisms—that evolves. The members of a species form groups, or **populations,** that occupy a particular region. Some species consist of just a single population living in one area, such as a small lake or an island. Other species are made up of more than one population, each in a different locality. For evolution to take place, change must occur in the genes that are present in the members of a population. These changes are passed on to the next generation during reproduction and are spread throughout the population by interbreeding. In order to understand this process, biologists have investigated the genetic changes in populations that generate evolutionary change.

You will recall from Chapter 12 that each individual receives one copy of a gene from each of its parents. Recall, also, that genes can occur in different forms, or alleles. Most alleles are either dominant or recessive, and an individual can be either homozygous (two identical alleles) or heterozygous (two different alleles) at any particular gene locus. Not all individuals that make up a population have the same alleles; consequently, there is *genetic variation* in the population. In humans, this variety is reflected by differences in pigmentation, facial characteristics, and blood types among different individuals and ethnic populations (Figure 33-2).

The sum of all the various alleles of all the genes in all of the individuals that make up a population is called the population's **gene pool.** If we could count every allele of every gene in every individual of a population, we could measure the genetic variation in a gene pool. The *relative* occurrence of an allele in the gene pool is expressed as an **allele frequency.**

To illustrate how allele frequencies are determined, we will examine a genetic trait that is prevalent in persons of central African descent: sickle cell anemia. Recall from Chapter 4 that sickle cell anemia is caused by a mutation that results in a substitution of one amino acid for another in one type of polypeptide chain that makes up a hemoglobin molecule (page 81). The mutant allele is denoted as S, while the normal allele is denoted as A. Approximately eight out of every 100 African Americans are heterozygous carriers of sickle cell anemia (genotype SA), and one out of every 500 have sickle cell anemia (genotype SS). Consequently, out of every 500 African Americans, there will be an average of 40 carriers (with a total of 40 copies of the S allele) and one person with the disease (with two copies of the S allele). Thus, out of 500 African Americans, having 1,000 copies of the gene, there will be an average of 42 copies of the sickle cell allele and 958 copies of the normal allele. The frequency for the S allele is $^{42}/_{1,000}$ or 0.042 (4.2 percent). Conversely, the frequency for the normal A allele is 0.958 (95.8 percent). (Calculations of this type can be done more quickly using the equation described in Appendix C.)

While differences in allele frequency exist between different human races (as illustrated in Figure 33-2), genetic analysis of a large variety of traits indicates that the overall differences are remarkably small. In the words of Richard Lewontin, "If everyone on earth became extinct except for the Kikuyu of East Africa, about 85 percent of all human variability would still be present in the reconstituted species."

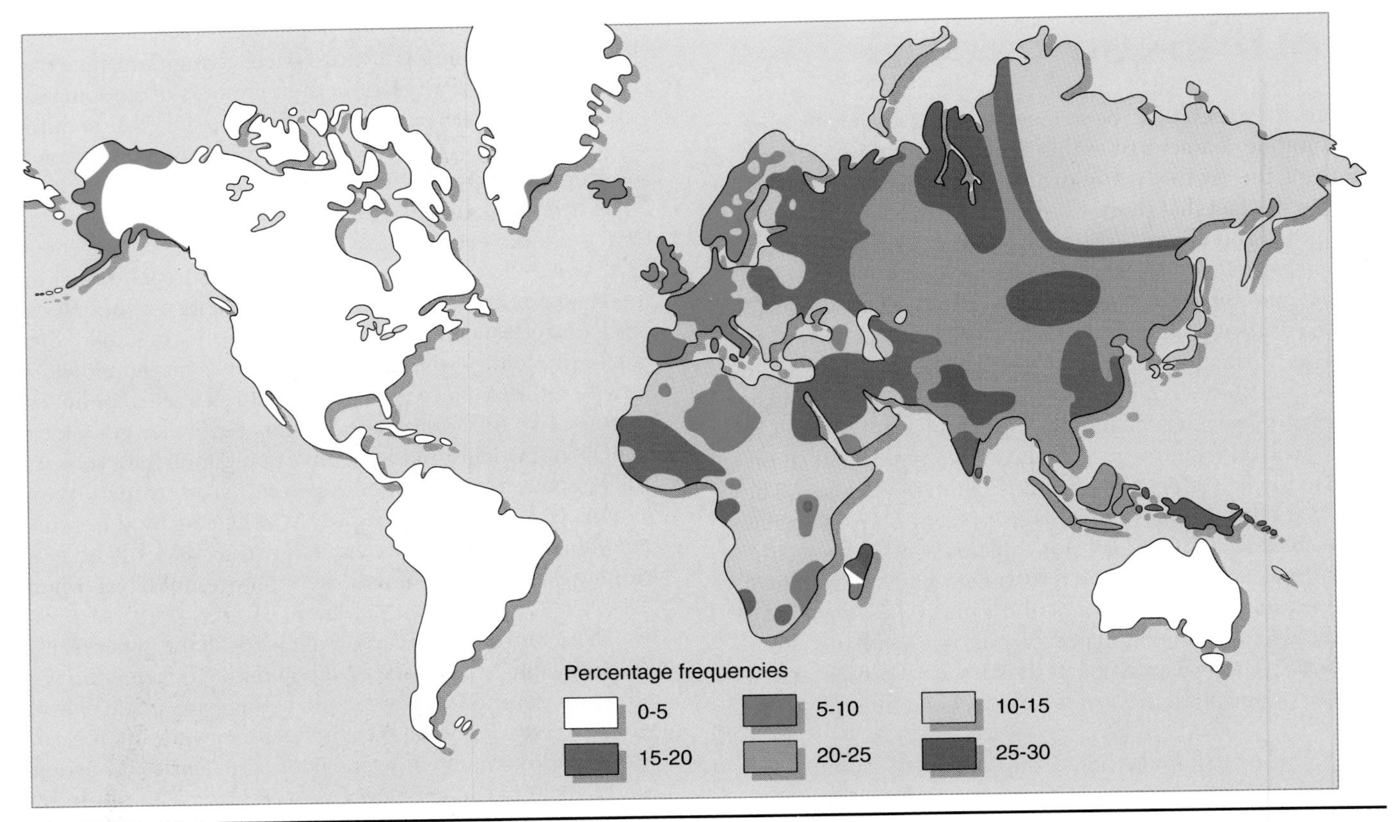

FIGURE 33-2
Frequency of the blood-type allele I^B in aboriginal (native) populations of the world. The frequency ranges from a high of about 30 percent (0.3) in Northern India and central Asia to a low of less than 5 percent (0.05) among American Indians and aboriginal Australians. From: *Modern Genetics,* 2nd ed. by: Ayala and Kiger. Copyright © 1984 by The Benjamin/Cummings Publishing Company. Reprinted by permission.

FACTORS THAT CAUSE GENE FREQUENCIES TO CHANGE OVER TIME

Evolution occurs when the composition of the gene pool changes. Therefore, a basic component of the evolutionary process is the change of allele frequencies over time. What causes allele frequencies to change? Sometimes the easiest way to understand a process is first to construct an artificial system or model in which the process does *not* occur. In this case, by uncovering the conditions that are necessary to keep allele frequencies *constant,* we automatically learn what forces will cause them to change.

In 1908, the British mathematician G. H. Hardy and the German biologist W. Weinberg independently discovered that under certain ideal conditions, allele frequencies will remain constant from generation to generation in sexually reproducing populations. Their demonstration is now known as the **Hardy-Weinberg Law** and is discussed in detail in Appendix C. Populations that are not changing—that is, that have the same allele frequencies from one generation to the next—are said to be at *Hardy-Weinberg equilibrium,* or **genetic equilibrium.**

Five "ideal conditions" must exist if a population is to remain at genetic equilibrium:

1. There must be an absence of mutation so that no new alleles appear in the population.
2. Individuals cannot migrate into or out of the population so that no new alleles enter, or existing alleles leave, the population.
3. The population must be very large so that it is not affected by *random* changes in allele frequency.
4. All individuals in the population must have an equal chance of survival; that is, there are no genetic traits that give individuals a survival advantage.
5. Mating must combine genotypes at random; that is, no preference is shown in the selection of a mate.

Based on these five conditions, we can identify those factors that disrupt genetic equilibrium and cause changes in the frequency of alleles in a population's gene pool:

1. *Mutation:* randomly produced inheritable changes in DNA that introduce new alleles (or new genes, as the result of chromosome rearrangements) into a gene pool.
2. *Gene flow:* the addition or removal of alleles when individuals exit or enter a population from another locality.
3. *Genetic drift:* random changes in allele frequency that occur solely by chance.
4. *Natural selection:* increased reproduction of individuals that have phenotypes that make them better suited to survive and reproduce in a particular environment.
5. *Nonrandom mating:* Increased reproduction of individuals that have phenotypes that make them more likely to be selected as mates.

These five forces, alone or in combination, determine the course and rate of evolutionary change. The role of each agent is slightly different but, in general, mutation and gene flow introduce new genetic material into a population, while genetic drift, natural selection, and nonrandom mating determine which alleles will be passed on to the next generation.

Mutation: The Source of New Alleles

A mutation is a random change in the DNA of an organism (page 286). Mutations occur spontaneously in all the cells of the body, but only those that occur in germ cells contribute to evolutionary change because only these cells can become gametes and pass the mutation on to the next generation. Mutations add new alleles to the gene pool, supplying the genetic foundation on which the other evolutionary forces operate.

When we see how quickly some insects have developed resistance to pesticides, such as DDT, it is tempting to propose that the appropriate mutations were stimulated to arise when the insects were *first* exposed to the pesticide, as a direct response to a change in the environment. This is not the case, however; mutations are random and unpredictable.

At any point in time, some mutations are beneficial, some are detrimental, and others are "neutral" and have no apparent effect on the survival or reproductive capacity of an organism. Many harmful mutations are immediately removed from the gene pool because they disrupt the structure and function of a protein whose activity is required for life to continue. Individuals with such lethal mutations typically die during embryonic development. (For example, many human zygotes fail to develop because of lethal mutations.) Other harmful mutations are masked by a dominant allele. For example, each of us is believed to carry an average of 7 to 8 lethal recessive genes. The fact that we are alive testifies to the role of the dominant allele on the homologous chromosome.

Whether a mutation is beneficial, detrimental, or neutral often depends on the environment in which the organism is living at the time. If the environment changes, the effects of the mutation on survival and reproduction can also change. For example, a mutation that causes an enzyme to function optimally at a higher temperature will be beneficial if the environmental temperature rises and will be detrimental if the temperature falls. It is likely that resistance to DDT was originally a neutral, or perhaps mildly beneficial, mutation that spread in low numbers throughout the insect populations through interbreeding. Only when the pesticide was sprayed did the DDT-resistance allele provide a strong survival advantage to the individuals that possessed it. Had no such allele already been present in the population when the individuals were exposed to the pesticide, no insects would have survived, and the population would have been wiped out. In fact, many bird species have been unable to adapt to the presence of the pesticide because the appropriate mutation is not present in their population. As humans continue to modify the earth's environments, it is important to remember that there is no guarantee that organisms will be able to adapt to environmental change.

Gene Flow: Exchanges of Alleles between Populations

It is common for animals or their larvae to migrate over large distances and for the seeds and pollen of plants to be dispersed by the wind or carried by birds to distant locations. Consequently, individuals from one population of a species are moved to another population, creating the opportunity for the transfer of alleles from one population's gene pool to another. The transfer of alleles between populations through interbreeding is called **gene flow.** Immigrants into a population may add new alleles to the population's gene pool, or they may change the frequencies of alleles that are already present. Emigrants out of a population may completely remove alleles, or they may reduce the frequencies of alleles in the remaining pool.

The amount of gene flow between populations varies greatly, depending on a number of factors, including the number of migrating individuals, the ease of movement, the harshness of the environment to be traversed, and the amount of interbreeding that actually takes place when migrants come in contact with a new population.

As described in the chapter opening section, gene flow is one of the factors responsible for the widespread resistance among insects to pesticides. Resistant individuals from one population emigrate into new populations, spreading resistance-conferring alleles into new geographic areas. The importance of gene flow can also be illustrated in humans. The fact that 70 percent of the alleles for cystic fibrosis in the United States can be traced to a single northern European (page 342) reveals how the influx of alleles can affect the genetic composition of a human population.

Genetic Drift: Random Changes in the Gene Pool

Genetic drift is a change in allele frequency that results simply by chance. Chance can affect allele frequency in several ways, but it is especially important during genetic recombination. When gametes are formed by meiosis, the segregation of chromosomes into any particular egg or sperm occurs by chance. When mating takes place, a great many of the gametes are wasted. Only a few happen to combine to form new individuals, representing a random sample of the parents' genes. Genetic drift may be caused by the spread or removal of alleles due to chance segregation into gametes that happen to participate in formation of offspring.

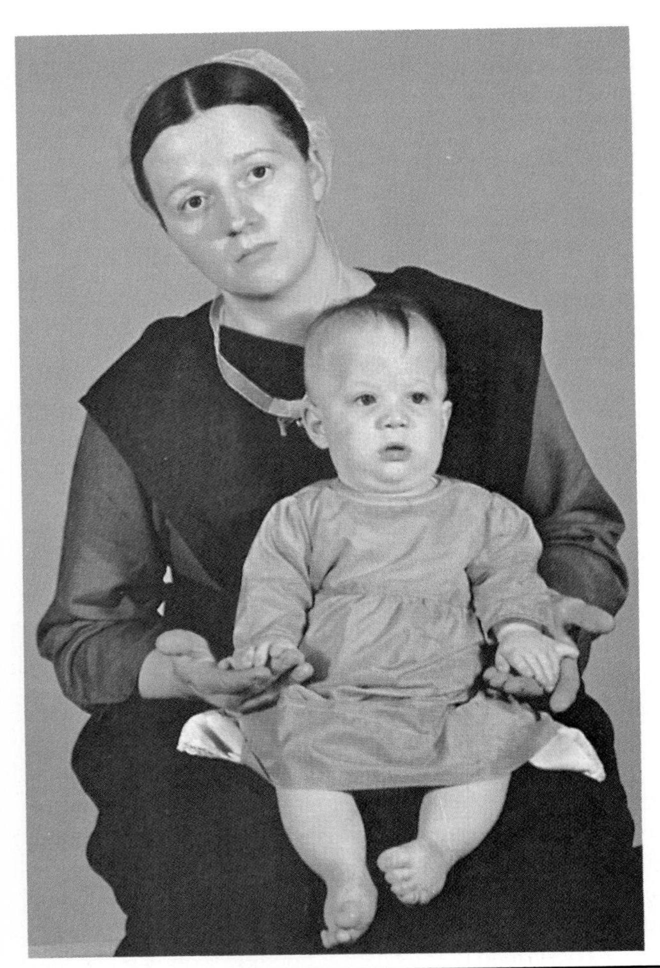

FIGURE 33-3
The founder effect. This Amish woman and her child are descendants of a small group of founding families who immigrated to Pennsylvania in the mid-eighteenth century. As a result of intermarriage between individuals within the community, a recessive allele for Ellis-van Creveld syndrome, which was present in one of the founders, has been able to pair with the same allele, producing homozygotes with the disorder. The child pictured here has the shortened limbs and extra fingers that characterize this syndrome.

Genetic drift occurs in populations of all sizes, but the effects of genetic drift are much more pronounced in small populations; in this case, the genetic composition of a few individuals has a significant impact on the gene pool. In large populations, chance effects tend to be averaged out. The same is true when flipping a coin. It is not unlikely that you will come up with heads or tails 75 percent, or even 100 percent, of the time if you flip a coin only four times, but the chance of this happening if you flip the coin 100 times is very remote. The chance becomes infinitesimally small if you flip the coin 1,000 times.

Even species that normally have large populations may pass through occasional periods when only a small number of individuals survive. During these so-called population **bottlenecks,** allele frequencies can change dramatically due to chance. Climatic changes, disease, predation, and natural catastrophes may reduce the size of a species to a very small number in a small area. During the last Ice Age, for example, the southward movement of glaciers in North America and Europe squeezed many plant and animal species into small areas, reducing population sizes to very low levels.

In a small population with only a few breeding individuals, complete mixing of the gene pool is possible. To take an extreme case, if there are only four individuals in a population, one of whom possesses a unique but selectively neutral trait, such as a dimpled chin, there is a good chance that this trait will be able to spread through the population in just a few generations. It is much less likely that the same trait will spread through a large population of thousands of individuals by chance alone.

An interesting account of genetic drift is provided by a study of the fishes that occupy warm springs in the Death Valley region of California and Nevada. Remarkably, one species, *Cyprinodon diabolis*, is completely confined to a single spring in Nevada, called Devil's Hole. This spring was formed about 12,000 years ago, after the close of the last continental glaciation, when the region was covered by a large lake. Devil's Hole is over 60 meters (200 feet) deep, but the fish are largely confined to a shallow shelf about 20 meters deep, giving these animals the smallest known range (area of occurrence) of any vertebrate species. In fact, it is possible for every fish to be in view at the same time. The number of individuals in the entire species population varies over time, but it is often as low as 50. Because of the small population size, random genetic drift is believed to have been important in the evolution of this species, resulting in a fish that is very different from its relatives.

The Founder Effect When a species expands into another region, a new population may be started by a small number of pioneering individuals. The founders are not likely to possess all the alleles found in the original parental population; even if they do, the proportion of each allele is likely to be different from that of the original population. Since the pioneers represent a small number of individuals, the new population that develops is likely to be strongly affected by genetic drift. This phenomenon is known as the **founder effect.**

Many examples of the founder effect are seen in isolated locations, particularly on oceanic islands of relatively recent geologic origin. When new islands appear, they tend to become colonized by a few members of a species that arrive on the island by chance and are affected by genetic drift. Among animals, a single female arriving on an island carrying fertilized eggs or embryos is all that is required to found a new population. Since many plants can reproduce by either self-fertilization or asexual reproduction, a single seed can colonize a new environment by itself. Consequently, the founder effect has been prominent in plant evolution. As the following example illustrates, the founder effect has also been documented in studies of human populations.

In the 1770s, a small number of Germans of the Amish sect emigrated to the United States and founded a community in Lancaster, Pennsylvania. For over 200 years, this population has remained, for the most part, reproductively isolated, with little intermarrying. One of the members of this founding group apparently carried a recessive allele for a rare form of dwarfism and polydactylism (extra fingers and toes), called the Ellis-van Creveld syndrome (Figure 33-3). A study carried out in the 1960s revealed that of the approximately 8,000 Amish living in the Lancaster area, 43 individuals were homozygous recessive for this allele and exhibited Ellis-van Creveld syndrome, representing more cases of this disorder than in the rest of the world combined! This study provides dramatic evidence of how the founder effect can generate populations whose allelic frequencies may be very different from those of the original population from which the founders arose.

Natural Selection: The Driving Force behind Adaptation

As a young man on the *Beagle,* Charles Darwin became convinced that organisms evolve over time. It wasn't until years later that he conceived of a mechanism that could actually cause that change. This mechanism was **natural selection.**

Ironically, one of the key observations that led Darwin to his conclusions about natural selection came from the practice of *artificial selection,* which has been used for thousands of years by plant and animal breeders to produce strains of crop plants and domestic animals. In this practice, offspring with desirable traits are selected from each generation for breeding purposes, while offspring lacking such traits are prevented from reproducing. The breeder continues to select in a particular direction, generation after generation, until he or she obtains the desired results. This practice often produces varieties of individuals that differ significantly from the original breeding stock. One of the most extreme examples of artificial selection can be seen in dog breeds, all of which are derived from an animal similar to the modern wolf (Figure 33-4*a*). In a few thousand years, artificial selection has produced varieties as different as the two pictured in Figure 33-4*b*.

Artificial selection demonstrated to Darwin and his contemporaries that continued selection was powerful enough to bring about large-scale changes within a species.

(a)

(b)

FIGURE 33-4

Products of artificial selection. ***(a)*** The wolf, *Canis lupus,* is the wild ancestor of the domestic dog ***(b)***. After generations of artificially selecting the traits they want to emphasize, breeders have produced dogs as different as the pair depicted here. Despite these differences, all domestic dogs are members of the same species and are capable of interbreeding to produce viable offspring.

The idea that *natural* selection could produce similar changes in *natural* environments therefore seemed a reasonable hypothesis.

Recall from Chapter 1 that, according to Darwin's theory of evolution by natural selection, not all individuals survive and reproduce equally well in a given environment. Therefore, some individuals contribute more offspring to the next generation than do others. As generations pass, those individuals with adaptive characteristics will become more common, and those with detrimental characteristics will be eliminated. In other words, the environment plays the role of the breeder in natural selection. Of all the forces that influence evolution, only natural selection generates populations whose members are better adapted to their environment.

In the opening pages of this book, we described a well-documented example of natural selection: the change in coloration of peppered moths *(Biston betularia)* that occurred during the Industrial Revolution in England. Examination of insect collections in museums show that prior to the 1850s, most members of the species living in English industrial areas were light to intermediate colored and delicately camouflaged to match the lichens that grew on the trees and rocks. In the latter half of the nineteenth century,

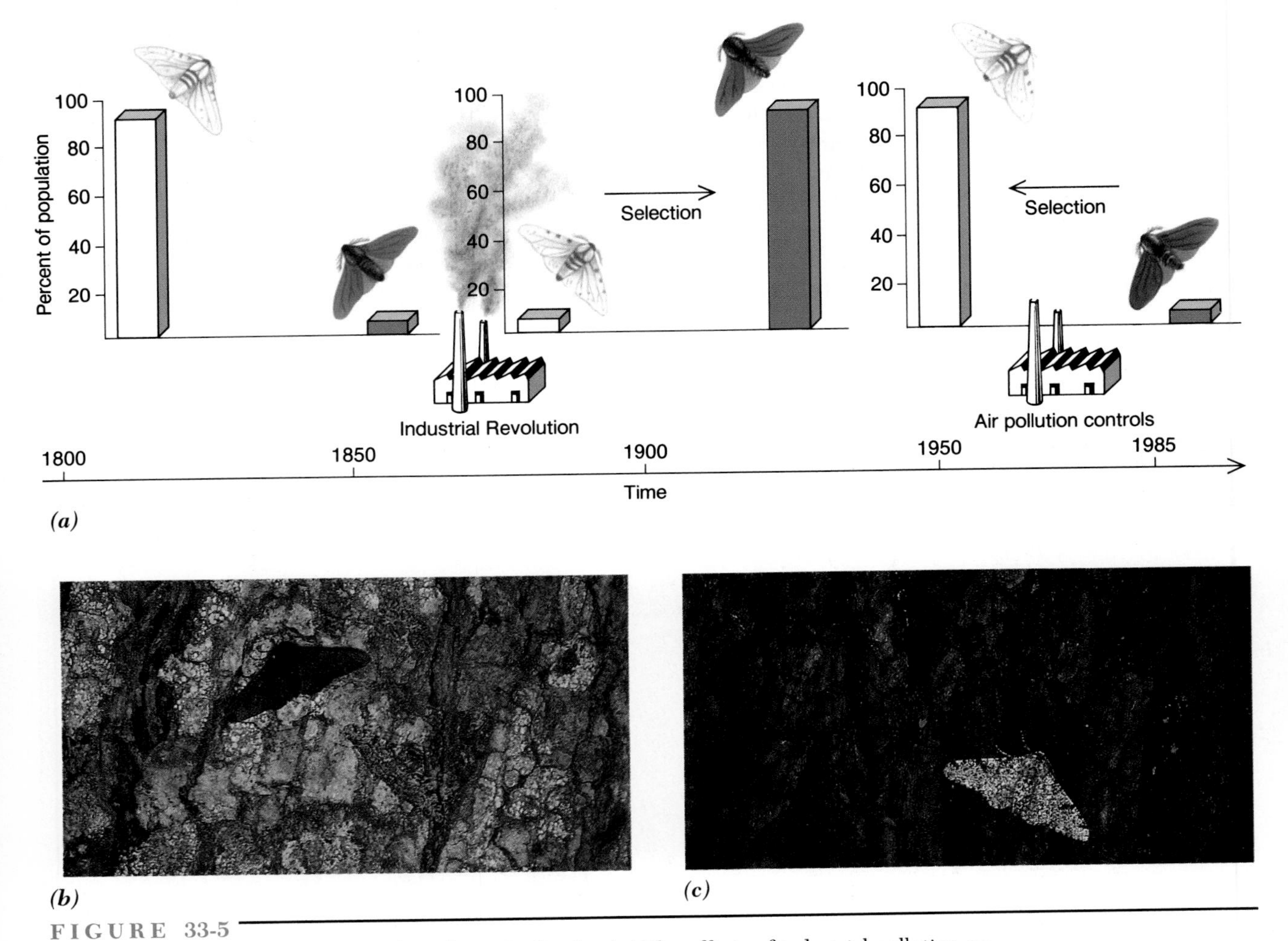

FIGURE 33-5
Natural selection of the peppered moth in England. ***(a)*** The effects of industrial pollution on the prevalence of speckled and dark varieties of peppered moths. Before the Industrial Revolution, speckled moths were more abundant than dark moths because the darker variety was more easily spotted on light, lichen-covered tree trunks ***(b)***. During the latter half of the nineteenth century, dark moths became more prevalent because they blended with soot-covered trees ***(c)***. Following recent pollution controls, light speckled individuals are once again more numerous than dark individuals.

TABLE 33-1

COMPARISON OF NUMBER OF RELEASED AND RECAPTURED SPECKLED VERSUS DARK PEPPERED MOTHS IN POLLUTED AND POLLUTION-FREE AREAS

Location	*Number Released*		*Number Recaptured*[a]	
	Speckled	*Dark*	*Speckled*	*Dark*
Birmingham (polluted)	64	154	16(25.0%)	82(53.2%)
Dorset (pollution-free)	393	406	54(13.7%)	19(4.7%)

[a] Percent recaptured is in parentheses.

booming industrial cities released tons of black soot from coal-burning factories, blackening nearby tree trunks and rocks and killing off the lichens. Light-colored moths were no longer protected from predation by birds. Instead, the dark-colored variety of peppered moth became better adapted to the new environment. By 1898, the dark moths made up 98 percent of the population of peppered moths near the industrial city of Manchester (Figure 33-5).

The effects of industrial soot on the frequency of speckled and dark peppered moths were verified experimentally in a study by H. B. D. Kettlewell in the 1950s. Both speckled and pigmented peppered moths were marked with a spot of paint under their wings and were released into both polluted and pollution-free areas. When the survivors were recaptured, the results confirmed that speckled moths were favored in pollution-free areas and that dark moths were favored in polluted areas (Table 33-1). As Darwin had originally suggested, it is the environment that selects which variants in a population will survive to reproduce.

Another example of natural selection comes from studies of sickle cell anemia in human populations. We noted earlier that the frequency of the sickle cell allele (S) in African Americans is about 0.042 (or 4.2 percent). The frequency of this allele in individuals living in central Africa is much higher (Figure 33-6), averaging about 0.12. From this value, we can calculate that approximately 1 percent of the inhabitants of this area will be SS homozygotes and stricken with sickle cell anemia (see Appendix C).

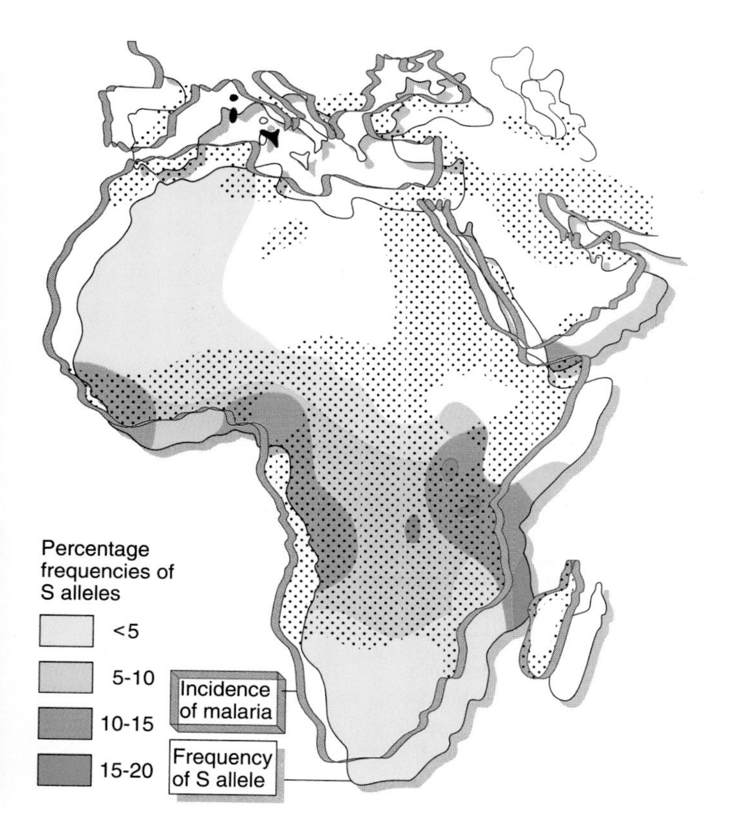

FIGURE 33-6

High levels of sickle cell anemia and malaria are found in the same geographic region. Upper map shows distribution of a virulent form of malaria caused by the protozoan *Plasmodium falciparum*. Lower map shows distribution of the sickle cell anemia allele (S) on the African continent. Because SA heterozygotes are more resistant to malaria than are AA homozygotes, the S allele has been selected for and kept at high frequencies in the African population. From: *Modern Genetics*, 2nd ed. by: Ayala and Kiger. Copyright © 1984 by The Benjamin/Cummings Publishing Company. Reprinted by permission.

How is it that so many central Africans have a disease that almost always causes them to die at an early age, before they can reproduce and pass their genes on to the next generation? Once it became apparent that the geographic distribution of sickle cell anemia in Africa was very similar to the distribution of a particularly lethal form of malaria, an answer to this question was suggested. Studies indicated that the blood cells of SA heterozygotes are much more resistant to the malarial parasite than are the blood cells of individuals who are homozygous for the normal allele (AA). Since malaria is very common in central Africa (Figure 33-6), heterozygotes are favored by natural selection over both types of homozygotes; homozygotes that possess two copies of the normal allele are more likely to die of malaria, while homozygotes with two copies of the sickle cell allele are very likely to die of sickle cell anemia before reaching reproductive age. As a result, heterozygotes pass the S form of the gene on to their offspring, maintaining the high level of an allele that would otherwise occur at a very low frequency in the population. The frequency of the S allele has dropped over time in the African American population, where this allele provides no selective advantage.

Patterns of Natural Selection Natural selection changes the frequency of certain phenotypes in a population. Three different patterns of change—stabilizing selection, directional selection, and disruptive selection—are recognized (Figure 33-7). All three patterns of natural selection may be acting on any species at any point in time.

Stabilizing selection occurs when individuals with extreme characteristics die or fail to reproduce, resulting in populations of individuals that possess intermediate characteristics (Figure 33-7*a*). For example, in most mammalian species, average birth weights tend to remain constant. Even in humans, until recent medical intervention, mortality was greater in human babies with high birth weights (due to difficulties in delivery) and low birth weights (due to decreased infant survival) than for babies of intermediate

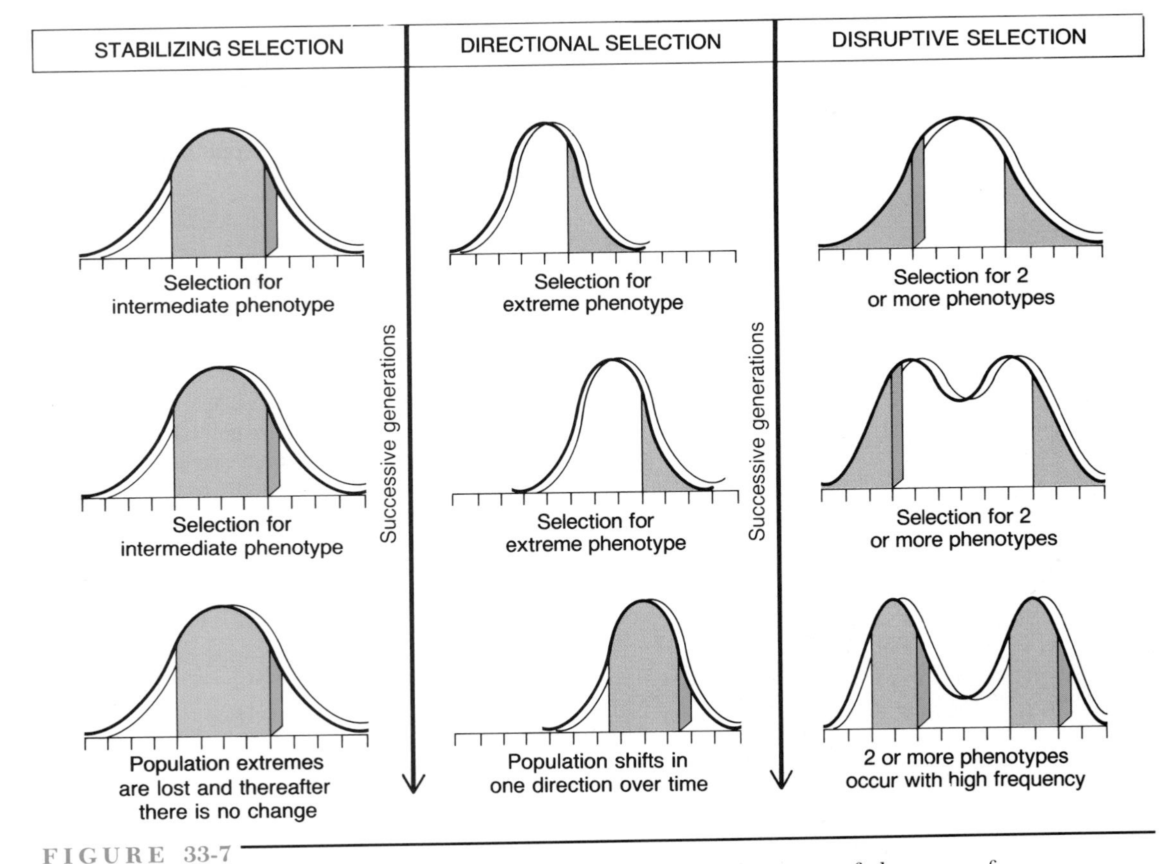

FIGURE 33-7
Three patterns of natural selection. The shaded areas represent the ranges of phenotypes for a particular trait that is favored by natural selection. A selection of the average (intermediate) phenotype ***(a)*** produces no change over three generations during stabilizing selection, whereas selection of extreme phenotypes leads to shifts in the characteristics of a population during directional ***(b)*** and disruptive selection ***(c)***.

weight. Since only the alleles for intermediate body weight were selected for, baby size remained relatively constant from generation to generation.

Stabilizing selection is most common in unchanging environments, where most populations achieve phenotypes that are optimally adapted to their surroundings. Over time, phenotypes that are closer to an optimum are retained, and extreme phenotypes are lost. This may explain why so-called living fossils, such as the ginkgo (maidenhair tree), chambered nautilus, and horseshoe crab (Figure 33-8), which inhabit stable environments with few competitors, have remained essentially unchanged for tens or hundreds of millions of years.

In **directional selection,** phenotypes at one extreme die or fail to reproduce, while those at the other extreme leave a higher number of offspring. When a phenotype at

(a)

(b)

(c)

FIGURE 33-8

"Living fossils." Certain organisms have remained virtually unchanged after many millions of years of stabilizing selection: ***(a)*** maidenhair tree *(Gingko)*, essentially unchanged for 100 million years; ***(b)*** chambered nautilus *(Nautilus)*, essentially unchanged for 180 million years; ***(c)*** horseshoe crab *(Limulus)*, essentially unchanged for 360 million years. (Inset: 145-million-year-old fossil horseshoe crab from Germany.)

one extreme is repeatedly selected, the frequency distribution gradually shifts in the direction of the favored phenotype (Figure 33-7*b*). Directional selection occurs when there is a change in the environment such that the phenotype at one extreme loses its selective advantage, while individuals possessing the phenotype at the other extreme increasingly survive and reproduce. We have already discussed two examples of directional selection in this chapter: the shift in frequency from the light-colored peppered moth to the dark form during the Industrial Revolution in England, and the increased resistance of mosquitoes to DDT. During human evolution, increased brain size and loss of body hair represent directional changes in phenotype.

In **disruptive selection,** extreme phenotypes become more frequent from generation to generation because indi-

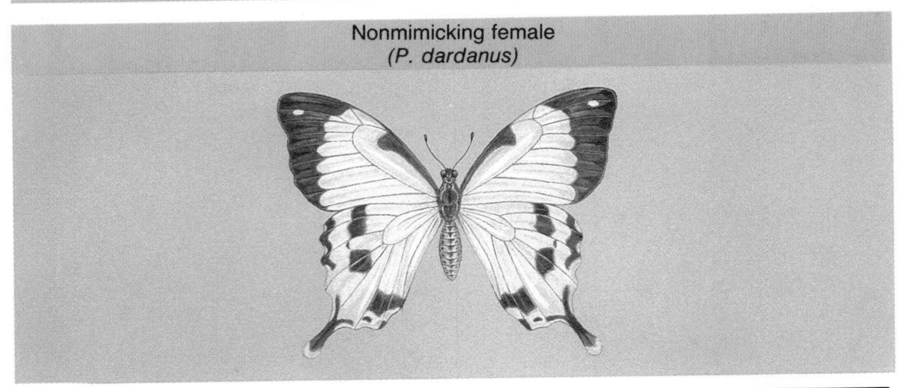

FIGURE 33-9
Altered states. Female African swallowtail butterflies mimic the appearance of local, foul-tasting butterfly species, creating strikingly different female phenotypes, even though they all belong to the same species. These different forms of females provide an example of disruptive selection and polymorphism.

(a)

(b)

FIGURE 33-10

Sexual selection. Although they may actually hinder the individual's mobility, brilliant displays of feathers ***(a)*** or racks of antlers ***(b)*** are products of sexual selection for characteristics that increase an individual's chances of mating.

viduals with intermediate phenotypes die or fail to reproduce (Figure 33-7*c*). Disruptive selection promotes **dimorphism** (two forms of a trait) or even **polymorphism** (several forms of a trait) in a population. This may happen in a diverse or cyclically changing habitat, where different individuals are adapted at different times or in different parts of the environment.

An example of disruptive selection is found in female African swallowtail butterflies *(Papilio dardanus)*. Although these butterflies are all members of the same species, the species is widespread, and individuals from one locale are strikingly different in appearance (phenotype) from those of other areas (Figure 33-9). Why would this occur? Some species of butterflies combat predation by concentrating noxious chemicals in their bodies from the plants on which they feed. After one or two nauseating bites, birds learn to recognize these distasteful butterflies and leave them alone. The female African swallowtail butterflies lack these chemicals and would make a tasty meal for a bird, were it not for the fact that they closely resemble (mimic) distasteful species. Distasteful butterflies tend to live in small populations, however. Female African swallowtails will only be protected by mimicking the species of distasteful butterfly that lives in their own small geographic area. Consequently, natural selection has favored the evolution of several distinct color patterns. Intermediate phenotypes between two local groups would not resemble any distasteful butterflies and would be devoured by the birds.

Nonrandom Mating

When individuals choose mates on the basis of their phenotypes, **nonrandom mating** occurs. Nonrandom mating can be caused by a number of factors. It frequently occurs when there is a preference for a particular type of mate or when the population becomes so small that there is no choice except to mate with a close relative.

Sexual Selection Not all characteristics favored by natural selection improve an individual's chances of *survival;* rather, some increase its chance of *reproducing.* The spectacular tail feathers of a peacock and the spreading antlers of a male deer (Figure 33-10) appear as if they could actually impede the animal's pursuit of food and escape from predators. Since these characteristics improve the chances of attracting females and reproducing, however (and in natural selection, passing on your genes is all that matters), they will be strongly selected for. This form of natural selection is called **sexual selection.**

Sexual selection often leads to *sexual dimorphism;* that is, visible differences between the male and female of the

BIOLINE
A Gallery of Remarkable Adaptations

The most striking result of evolution by natural selection is adaptation, the ways in which organisms seem to fit exactly with the world in which they live. Natural selection has resulted in some truly remarkable adaptations, some of which are morphological, such as sharp teeth and claws, horns, and trichomes. As the following examples illustrate, however, adaptations can also be behavioral or reproductive.

a. *A fishy lure—a morphological adaptation.* The scorpion decoy fish *(Tricundus signifer)* has a dorsal fin that resembles a smaller fish, complete with its own "dorsal fin," "eye," and "mouth." The scorpion fish swishes its dorsal fin back and forth, luring small, would-be predators. Within a tenth of a second, the hopeful diner quickly becomes the dinner, fatally fooled by a very artful angler.

b. *Disguise—a morphological adaptation.* The tiger swallowtail butterfly *(Papilio glaucus)* progresses through a series of larval and pupal stages, each with its own deceptive morphological adaptations. The first larval stage resembles bird droppings. (What predator would eat that?) Three stages later (photo), the green larval caterpillar blends in with the leaves it eats. The caterpillar also has large, false "eyes" that frighten away predators. The pupal stage masquerades as a broken

same species (Figure 33-11). Sexual selection is common among animals because a female's reproductive success is limited by the number of eggs she can produce in her lifetime, and a male's reproductive success is limited by the number of females he can inseminate. Therefore, it is to the female's advantage to choose the most fit male as her mate, and it is to the male's advantage to attract as many females as possible. This leads to natural selection of certain male characteristics, either through male competition with one another or through female choice. On the one hand, males may compete directly by fighting, or they may compete for territory, the possession of which attracts females. Consequently, in these species, males develop characteristics that enable the animal to fight or intimidate other males, like the antlers of a deer or the huge body size of the male elephant seal. On the other hand, females choose a mate, so natural selection favors those characteristics that females prefer. As a result, characteristics such as the bright-colored plumage of male birds, which is perhaps best exemplified by the gaudy tail feathers of peacocks, become exaggerated.

FIGURE 33-11

Sexual dimorphism in elephant seals and wood ducks. The larger elephant seal and the more brightly colored duck are the males.

twig on a tree trunk, camouflaged from hungry predators.

c. *Playing dead—a behavioral adaptation.* When threatened, the ringed snake *(Natrix natrix)* feigns death by dropping its head, dangling its tongue out of its mouth, and lying completely motionless. These actions help secure the snake's safety because most predators avoid dead organisms.

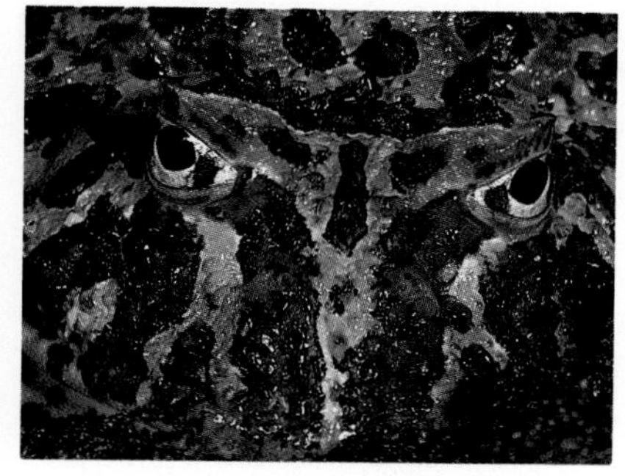

d. *Safe as the ground you walk on—a behavioral adaptation.* The camouflaged horned toad *(Ceratophrys ornata)* buries its body in mud, leaving only its eyes and large jaws protruding. Unwary prey quickly disappear as they move over the concealed head.

e. *Torpedo seeds—a reproductive adaptation.* The seeds of red mangrove trees germinate while they are on the tree. When released, the streamlined radicle slices through the water, planting the seedling upright. Seedlings that don't reach the bottom are able to float for months, until they run aground in shallow water and take root.

f. *Two in one—a reproductive adaptation.* Life as an independent organism is over almost as soon as it begins for the male deep-sea angler fish *(Edriolychunus schmidti).* As an adaption to allow members of the opposite sex to find each other in the blackness of the deep sea, the newly hatched male angler fish permanently attaches itself to the female by sinking its jaws into her body. The female's skin grows over the male's body, and the individuals' circulatory systems connect. The male becomes incorporated into the female body and is reduced to nothing more than a small sperm factory.

FIGURE 33-12
The cheetah population has lost most of its genetic variability due to inbreeding after the population was drastically reduced in size within the past 20,000 years.

Inbreeding Most harmful alleles originate as rare mutations and are limited to a small percentage of the population. Usually, these mutations exist as recessive alleles. If inbreeding does *not* occur, the chances are slight that two unrelated organisms will have the same harmful recessive alleles. Thus, it is unlikely that many offspring with the homozygous recessive phenotype will be generated, only to die of a genetic disease. If the two organisms are closely related and have received their alleles from a common ancestor, however, the chances of their carrying the same harmful recessive alleles are much greater. If these related organisms mate, chances are even greater that they will produce offspring with the defective phenotype. Obviously, when these offspring fail to reproduce and die, *all* of their genes (beneficial and deleterious) are removed from the gene pool. Therefore, since inbreeding increases the likelihood of death due to genetic defects, the population will lose more and more of its variability. If the environment changes, the population may lack sufficient variability to adapt to the change and may become extinct.

Smaller populations are affected more dramatically by nonrandom mating (and inbreeding, in particular) than are large ones. The effects of inbreeding in a small population are illustrated by the cheetah (Figure 33-12), the world's fastest land animal. At the present time, there are approximately 20,000 cheetahs left in the world, a number that normally would not indicate any danger of extinction. But cheetahs are different. Biologists who have studied these cats have discovered that cheetahs are not a healthy species. Cheetah cubs have a much lower survival rate than do the cubs of other species of large cats. Cheetah cubs are more susceptible to diseases, such as distemper, and the adult males typically produce a small number of sperm, most of which have an abnormal shape.

The reason for the poor health of cheetahs became apparent when a group of biologists led by Stephen O'Brien of the National Institutes of Health analyzed the gene pool of the species to find out why cheetahs were so difficult to breed in captivity. The group began by examining 50 different blood proteins from a variety of individuals, expecting to find a variety of allelic forms of these proteins. To the biologists' surprise, no differences were found in the proteins among the population; all of the cheetahs were homozygous for all of the genes coding for these proteins. Subsequent studies indicated that cheetahs are so genetically similar that they won't even reject tissue grafts from one another, a phenomenon unknown among other mammalian species (page 650).

What has happened to the cheetah? Several million years ago, cheetahs were found in abundant numbers among the African fauna and presumably exhibited a normal variety of alleles within their gene pool. Sometime within the past 20,000 years, conditions arose that apparently decimated the animal's numbers. As the population of cheetahs dwindled to a few individuals, the gene pool was drastically reduced. In other words, the cheetah went through a genetic bottleneck. As the few survivors interbred, the offspring became increasingly homozygous. Furthermore, harmful recessive mutations were paired more frequently by inbreeding, producing homozygous recessive individuals that died, taking with them some of the desirable alleles at other loci, further reducing genetic variability.

Today, the cheetah population probably lacks the variable phenotypes that are needed to ensure the survival of the species. If the cheetah population is to be maintained, new genetic variability will have to come from mutation; the alternative is extinction. The fate of the cheetah illustrates the genetic effects of extreme inbreeding. Virtually all human societies have taboos against incest—sexual relations between parents and their offspring or between brothers and sisters. Approximately half of the states in the United States have laws that prohibit marriages of first cousins. On the average, offspring from marriages between first cousins are about twice as likely to be born with a serious inherited disorder than are offspring from unrelated parents.

The examples we have discussed up to this point—pigmentation in peppered moths, sickle cell anemia in humans, loss of genetic variation in cheetahs, pesticide resistance in mosquitoes—are considered examples of **microevolution** because they result from changes in the allele frequency of a species' gene pool but they have not resulted in the appearance of new species, a phenomenon referred to as **macroevolution.** Microevolution reveals the process of evolutionary change over a short enough period of time so that it can be documented and studied. The occurrence of microevolution allows biologists to study the underlying mechanisms—mutation, gene flow, genetic drift, natural selection, and nonrandom mating—which, given sufficient time, lead to macroevolution, the subject of the remainder of this chapter.

Evolution is the greatest unifying concept in biology. Evolution explains why there are so many different types of organisms on earth; why each species is so well adapted to its particular habitat and lifestyle; and why a species shares many basic features with distant relatives while possessing unique features that distinguish it from all other species. (See CTQ #2.)

SPECIATION: THE ORIGIN OF SPECIES

Biologists have hypothesized that there are more than 5 million species of organisms alive today, even though only about 1.8 million have been described and named so far. Evidence indicates that all of these species, plus all those that lived in the past and have become extinct, descended from a single ancestor that lived approximately 3.5 billion years ago. **Speciation**—the process by which new species are formed—occurs when one population splits into separate populations that diverge genetically from one another to the point where they become separate species. Speciation has produced the millions of species that have inhabited the earth throughout time.

WHAT ARE SPECIES?

Species is a Latin word meaning "kind." Kinds, or species, of organisms were originally identified by their appearances because members of a species typically look alike. In some cases, identifying an individual as a member of a particular species is not so simple, however, because individuals from distinct species may appear very similar. Even among large, familiar vertebrates, such as giraffes, elephants, or camels, there are usually two or more species that closely resemble one another (Figure 33-13*a*). When we consider the more numerous smaller animals and plants, the problem of identification becomes even more apparent. To distinguish closely related species of insects, for example, a biologist may have to resort to the examination of microscopic bristles.

FIGURE 33-13
Are these animals members of the same or different species? ***(a)*** These two elephants, one from Africa (*Loxodonta africana*), the other from India (*Elephas maximus*), are members of two different species. The Indian elephant (right) has smaller ears and more pronounced "bumps" on its head. ***(b)*** All of these sea stars are members of the same species, even though they show marked differences in coloration.

Distinguishing between species of similar morphology may require the analysis of biochemical, ecological, and behavioral traits, as well as those visible to the eye. Some species may be difficult to identify because the members of the species population have different appearances due to a high degree of genetic variation. This is illustrated by the polymorphic African butterflies depicted in Figure 33-9. Similarly, one of the common sea stars of the Pacific coast occurs in a wide variety of colors (Figure 33-13*b*), yet all of these animals belong to a single species.

Why should two insects that differ slightly in bristle pattern be considered separate species, while two sea stars of totally different color (or two dogs of totally different body shape) are included in the same species? The most important criterion for defining a species is that members are capable of producing other members by mating within the community. One definition of species that incorporated this concept of shared reproduction was given by Ernst Mayr of Harvard University in 1940 and is now known as the **biological species concept.** According to this definition, "Species are groups of actually or potentially interbreeding natural populations which are reproductively isolated from other such groups." By including the phrase "actually or potentially," Mayr acknowledged that although distance, time, or geographic barriers may separate some individuals, the individuals are still members of the same species if they can interbreed once the barrier is removed.

Furthermore, the interbreeding must be natural. Individuals that do not normally breed in the wild are sometimes mated in captivity. For example, zoos sometimes display "tiglons" (offspring from the mating of a tiger and a lion) or similar hybrids. These animals do not occur in nature; thus, they have no real effect on the evolutionary history of groups. While Mayr's definition works well in defining animal species, it does not always hold for plants (see Steps to Discovery, Chapter 38). Among shrubs and trees, in particular, closely related species may interbreed and form fertile hybrids that then give rise to a population of hybrid individuals.

REPRODUCTIVE ISOLATING MECHANISMS

According to the biological species concept, members of one species are *reproductively isolated* from members of all other species. Accordingly, reproductive isolation, which prevents the exchange of genes between populations, is the first step leading to the formation of new species. Once the gene pools are isolated, the separated populations inevitably diverge because of differences in mutation, mating patterns, genetic drift, and natural selection. Over time, the isolated populations amass morphological, physiological, and behavioral differences that prevent them from interbreeding. Consequently, even if the original cause of isolation is removed, the populations remain reproductively isolated; they have become different species.

Barriers that prevent the exchange of alleles between populations (gene flow) are called **isolating mechanisms.** Isolating mechanisms are divided into two categories, depending on whether the isolation prevents a zygote from forming **(prezygotic isolating mechanisms)** or eliminates the success of such crosses as they occur **(postzygotic isolating mechanisms).** Examples of the two categories of isolating mechanisms are presented in Table 33-2.

PATHS OF SPECIATION

The millions of different species that exist today did not emerge by any single sequence of events but have come into existence by a number of different paths of speciation.

Phyletic Speciation

Darwin entitled his great book *On the Origin of Species,* but in it he discussed only how populations could change under the influence of natural selection, as opposed to the other mechanisms described above, such as gene flow and genetic drift. For Darwin, speciation was the simple, gradual accumulation of changes in a lineage through time, until the group was distinct enough to be considered a new species. This process is now called **phyletic speciation.** Although phyletic speciation undoubtedly occurs, speciation is more often the result of one coherent reproductive group splitting into two or more new, discrete species.

Allopatric Speciation

Allopatric speciation (*allo* = other, *patri* = habitat) is believed to be the most common type of speciation. Allopatric speciation typically occurs when a physical barrier, such as a mountain range, a river, or even an oil pipeline, geographically separates a population from its parental population, thereby cutting off gene flow between the two. While isolated, the separated population develops a number of genetic differences, including a reproductive barrier, that distinguish it from the main population. At this point, the two populations can be considered separate species. For example, the dozen or so species of finches discovered by Darwin on the Galapagos Islands are thought to have evolved as the result of their geographic separation from the parental species in Panama and from one another on different islands (see Figure 1-10). Another example of allopatric speciation is provided by several hundred species of fruit flies living on the Hawaiian Islands; all of these fruit flies are believed to have arisen from a single parental species.

Parapatric Speciation

Parapatric speciation (*para* = beside) is thought to occur in populations that lie adjacent to one another. Gene pools diverge because the environment varies sufficiently in the different locales. As a result, different traits are selected in each population. In one study of grasses growing in regions of abandoned mines, for example, investigators found that populations living in areas of toxic mine wastes had devel-

TABLE 33-2
ISOLATING MECHANISMS

1. ***Prezygotic Isolating Mechanisms***

Ecological Isolation Different habitat requirements separate groups, even though the inhabitants may exist in the same general location. Example: Head and body lice are morphologically very similar, yet they live in different "habitats" on a single human body. Head lice live and lay eggs in the hair on the head of a human, whereas body lice live and lay their eggs in clothing. Both suck blood for nutrition.

Geographical Isolation Emerging mountains, islands, rivers, lakes, oceans, moving glaciers, and other geographic barriers keep groups isolated. Example: Different tortoises are found on different Galapagos Islands; surrounding oceans keep tortoise populations isolated.

Seasonal Isolation Differences in breeding seasons prevent gene flow, even when populations are found in the same area. Example: Two populations of bigberry manzanitas grow close together in the mountains of southern California, yet the populations do not interbreed because one completes blooming 2 weeks before the other begins to bloom.

Mechanical Isolation Physical incompatibility of genitalia. Example: Genital structures differ in shape for alpine butterfly species, even though these butterflies look nearly identical in all other ways.

Behavioral Isolation Differences in mating behavior prevent reproduction. Example: Many animals have evolved complicated courtship activities before breeding. Some species of fruit flies *(Drosophila)* are indistinguishable to our eyes, yet they do not mate with each other because of differences in courtship behavior.

Gamete Isolation Sperm and egg are incompatible. Gamete isolation is a common isolating mechanism in many plant and animal species.

2. ***Postzygotic Isolating Mechanisms***

Hybrid Inviability Zygotes or embryos fail to reach reproductive maturity. Example: Hybrid embryos formed between two species of fruit flies fail to develop.

Hybrid Sterility Fertilization is successful between two species, but hybrid progeny are sterile. Example: A mule is a sterile hybrid produced from a mating between a horse and a donkey.

oped a tolerance to heavy metals which was not present in populations growing in adjacent, nonpolluted areas. In addition to differences in their sensitivity to metals, the two populations have diverged in time of flowering, resulting in reproductive isolation.

Sympatric Speciation

Sympatric speciation (*sym* = same) occurs in populations where individuals continue to live among one another, even though some type of *biological* difference, such as the time of the year when gonads mature, has divided the members into different reproductive groups. The best-accepted cases of sympatric speciation occur in plants as a result of **polyploidy**—an increase in the number of sets of chromosomes per cell.

The appearance of tetraploid (4N) offspring from diploid (2N) parents is not an uncommon occurrence among certain types of plants. Once formed, the tetraploid cannot interbreed with diploid members of the population because of chromosome incompatibility (page 260). Consequently, the tetraploid plant must either engage in fertilization with another tetraploid individual in the population, or it must produce a population of tetraploid plants on its own (either by self-fertilization or asexual reproduction) which can then interbreed. Regardless of the specific pathway, one interbreeding population is converted into two reproductively isolated populations, setting the stage for speciation. Since many plants are self-fertile and capable of asexual reproduction, polyploidy has been very important in plant evolution and speciation. More than 40 percent of flowering plant species living today are polyploids.

Hybridization

Rapid speciation by **hybridization** occurs when two distinct species come into contact, mate, and produce hybrid offspring that are often reproductively isolated from either parent but not from one another. In just one generation, an entirely new species can be generated by hybridization. At first glance, it may seem unlikely that hybridization could possibly produce viable offspring; in fact, viable offspring are rare in cases of animal hybridization. Plants are different, however, because they are more tolerant of polyploidy than are animals and because a single individual with a unique complement of chromosomes can generate a new population by self-fertilization or asexual reproduction.

Common wheat *(Triticum asestivum)* is believed to have evolved by hybridization. The original stock was probably similar to an ancient crop plant we now call einkorn wheat *(Triticum monococcum)*. This wheat appears to have been cross-pollinated by a wild grass *(Aegilops speltoides)* that grows abundantly on the edges of wheat fields in southwestern Asia. Each of these species has seven pairs of chromosomes, but their hybrid offspring have 14 pairs. This hybrid offspring is similar to what we now call emmer wheat *(Triticum durum)*. Subsequently, emmer wheat hybridized with goat grass *(Aegilops squarrosa)*, which has seven pairs of chromosomes and is found in the mediterranean area. The result is our modern species of wheat, which has 21 pairs of chromosomes.

Most new species are thought to have resulted from the separation of a population into two groups by a physical barrier that prevented interbreeding. The separated populations are influenced differently, causing them to diverge from one another and to become separate species. (See CTQ #4.)

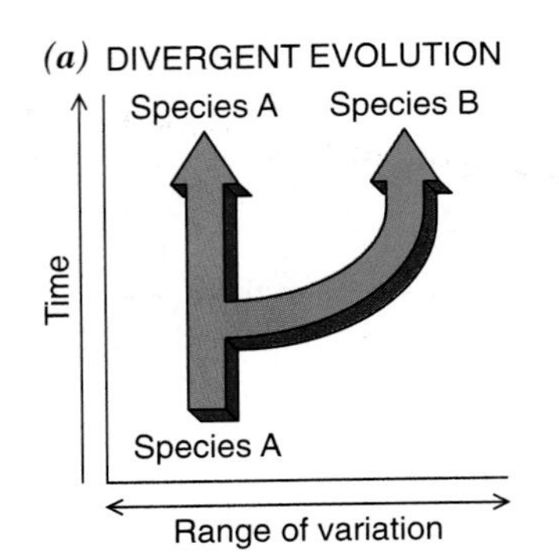

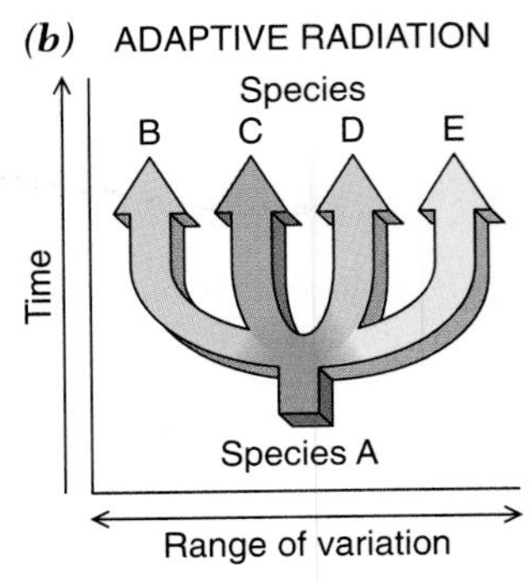

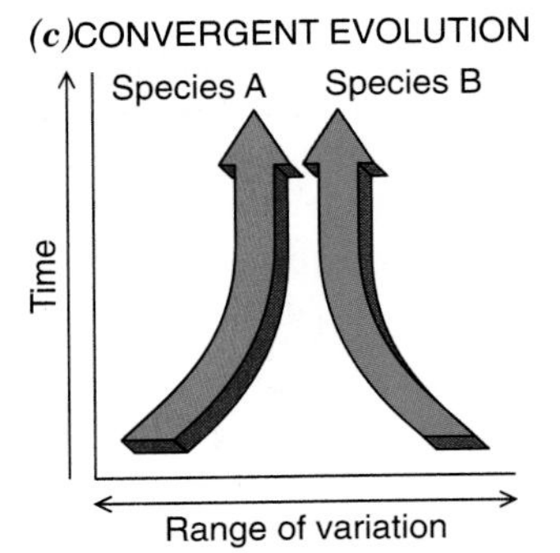

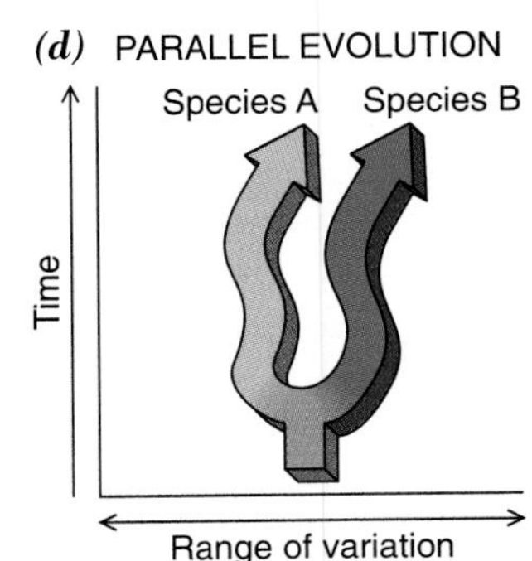

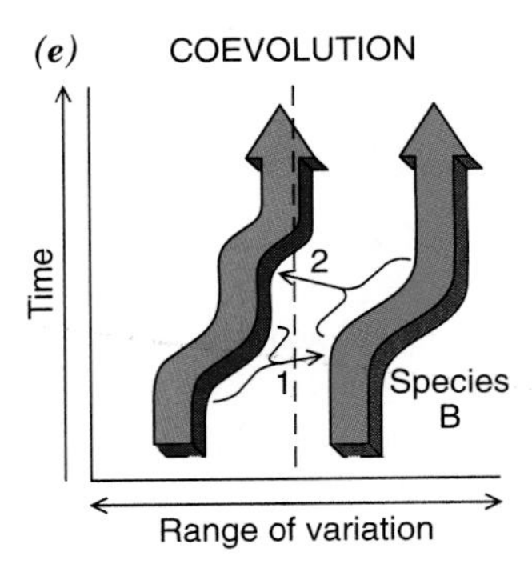

FIGURE 33-14

Patterns of evolutionary change. ***(a)*** Divergent evolution: One species splits into two species. ***(b)*** Adaptive radiation: One species gives rise to many new species that are adapted to different types of habitats and/or food sources. ***(c)*** Convergent evolution: Unrelated species evolve similar characteristics as the result of similar selective pressures. ***(d)*** Parallel evolution: Two related species remain similar over long periods of time. ***(e)*** Coevolution: Two species evolve in such a way so that changes in one causes reciprocal changes in the other.

PATTERNS OF EVOLUTION

As new species form and adapt to their environments through natural selection, different patterns of evolution may emerge. The most common pattern, **divergent evolution,** occurs when two or more species evolve from a common ancestor and then become increasingly different over time (Figure 33-14*a*). Divergent evolution forms the basis for phylogenetic branches, whereby one ancestral species gives rise to two distinct lines *(lineages)* of organisms that continue to diverge. Monkeys and apes, for example, diverged from a common ancestor, as did apes and humans (Chapter 34).

Sometimes, when members of a species move into a new area with many diverse environments, new species form and rapidly diverge, producing a variety of related species that are adapted to different habitats. This rapid divergent evolution is referred to as **adaptive radiation** (Figure 33-14*b*). One of the most astonishing examples of adaptive radiation occurred on the isolated continent of Australia, where the diversity of habitats and the absence of competitors sparked an adaptive radiation of marsupials (pouched mammals). Beginning with a small, opossum-like marsupial that lived about 100 million years ago, a diverse array of marsupials evolved, ranging from kangaroos that hop across the open Australian prairies to koalas that cling to branches in the forest's trees.

There are a limited number of solutions to any environmental problem. For example, rapid movement through

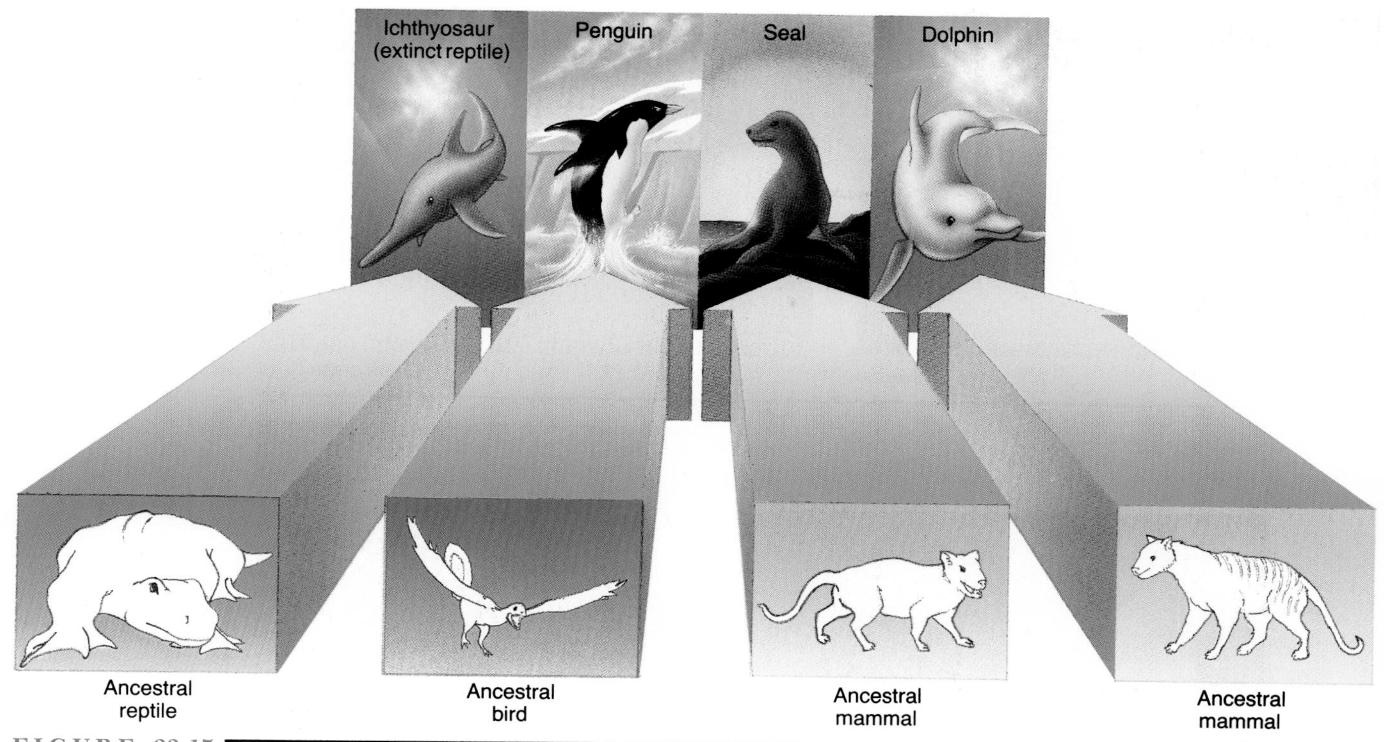

FIGURE 33-15

Convergent evolution. These marine animals are descended from different ancestors. They have all developed similar streamlined bodies and paddlelike front limbs that adapt them to life in the water.

the water requires a streamlined body shape, while movement through the air requires wings. Therefore, when species with different ancestors colonize similar habitats, they may independently acquire similar adaptations and resemble one another superficially. This phenomenon is called **convergent evolution** (33-14*c*). For example, each of the four marine animals depicted in Figure 33-15 is descended from a different terrestrial ancestor, but they all share certain common adaptations, such as streamlining and paddlelike forelimbs.

The similarity among many Australian marsupials and placental mammals on other continents is another example of convergent evolution. The Austrialian marsupials evolved independently from their placental counterparts and are only distantly related, yet the two groups include animals with strikingly similar characteristics. For example, there are both marsupial and placental "anteaters," "wolves," and "flying squirrels" (Figure 33-16). Fossils indicate that there was even a marsupial saber-toothed tiger.

Parallel evolution occurs when two species that have descended from the same ancestor remain similar over long periods of time because they independently acquire the same evolutionary adaptations (Figure 33-14*d*). Parallel evolution occurs when genetically related species adapt to similar environmental changes in similar ways. For example, the ancestral arthropod had a segmented body with a pair of legs on each segment. In all three major arthropod lineages that have descended from this ancestor (the crustaceans, insects, and spiders), the number of legs has decreased, and the body segments have become fused, forming larger structures with specialized functions.

Since organisms form part of the natural environment, they can also act as a selective force in the evolution of a species. In nature, species frequently interact so closely that evolutionary changes in one species may cause evolutionary adjustments in others. This evolutionary interaction between organisms is called **coevolution** (Figure 33-14*e*). Flowering plants and their insect pollinators have coevolved for millions of years, leading to many finely tuned structural and behavioral relationships between flowers and pollinators (Chapter 20). Another example of coevolution is found in predator–prey interactions, where improvements in the hunting ability of a predator favor the survival of a prey with characteristics that increase its ability to escape. Parasites and their hosts also coevolve. Over time, parasites tend to become less destructive of their hosts (a dead host means a dead parasite), and hosts tend to become more resistant to the parasite.

Over long periods of time, several different patterns of evolution may emerge. Most organisms living today are the result of an adaptive radiation whereby an ancestral species gave rise to a number of descendant lines containing organisms that are adapted to different habitats. (See CTQ #5.)

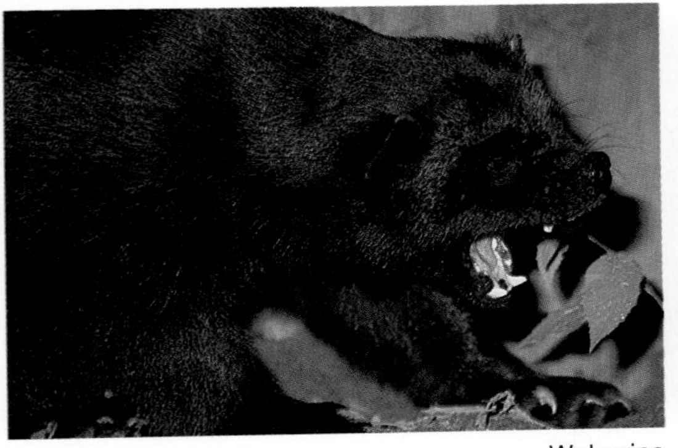

Wolverine

Tasmanian Devil

Southern Flying Squirrel

Sugar Glider

FIGURE 33-16
Living examples of convergent evolution. Australian marsupial mammals and placental mammals on different continents have similar features because they have adapted to similar habitats. Placental mammals are shown on the left, and the Australian marsupial counterpart is on the right.

EXTINCTION: THE LOSS OF SPECIES

Extinction, or the loss of a species, is an important part of evolutionary history. When you consider that approximately 99 percent of the species that have existed since the beginning of life on earth are no longer alive, it becomes apparent that extinction is the ultimate fate of most, if not all, species.

Extinction can occur gradually over a period of tens of thousands of years, or quite suddenly in just one to a few generations. Rapid extinctions are more common among organisms that live in small populations or in geographically restricted areas, such as a single lake or forest. One period of local drought or forest fire can mean the extinction of the entire species. At any point in time, extinction may be limited to just one or a few species, or it may involve the sudden, simultaneous extinction of a multitude of species in a **mass extinction** event (Chapter 35).

Species may become extinct when they lack genetic variability or when they find themselves in the wrong place at the wrong time. In other words, extinction is due to either bad genes or bad luck. In the first case, a species can become extinct when the environment changes and none of the species' members has the genetic makeup that will enable the organism to survive under the new conditions. In the second case, a species may face an unusual catastrophe that essentially eliminates all life in its habitat. Some of the causes that have been proposed for mass extinctions in the past include asteroid impact, volcanic eruptions, drastic changes in sea level, and radical shifts in the earth's climate. Today, organisms on earth are faced with a new cause of mass extinction: The unbridled destruction of natural habitats by humans has increased the extinction rate from a long-term average of about one species each 1,000 years to hundreds, and perhaps thousands, of species in a single year.

Most of the species living today have evolved in recent times. While some of the species that are no longer here were gradually transformed into other species by phyletic speciation, most were unable to adapt to changing conditions and became extinct. (See CTQ #6.)

THE PACE OF EVOLUTION

Darwin viewed evolution by natural selection as a steady, uninterrupted process. He believed that just as natural selection adapted a population to its environment, the process could also turn that population into a new species and eventually found a whole new order, class, or phylum. The discovery of the importance of mutation, gene flow, genetic drift, and nonrandom mating, as well as natural selection, seemed to confirm Darwin's view that most evolution occurs in small, adaptive steps. Under such a model, evolution proceeds by **gradualism** (Figure 33-17*a*). This view has been criticized by some paleontologists (biologists who study the fossil remains of animals that lived in the past), who contend that fossil evidence does not show a gradual succession of forms. Rather, the analysis of fossils of numerous groups indicates that long periods without significant change (periods of "stasis") are interspersed with short periods of very rapid change.

In 1972, paleontologists Niles Eldredge of the American Museum of Natural History in New York and Stephen Jay Gould of Harvard University proposed a hypothesis called **punctuated equilibrium** (Figure 33-17*b*) to explain this pattern of evolution. The punctuated equilibrium model includes two separate proposals. The first states that speciation, when it occurs, is a rapid process. We have already seen how allopatric speciation can lead to new species and how small populations are changed more rapidly than are larger ones. When both phenomena occur together (allopatric separation of small populations), spurts of speciation can occur.

The second proposal is that, once formed, species exist for long periods of time without change, unless the environment is altered in some way. This stasis occurs because, even in semistable environments, species often reach a population size large enough for stabilizing selection and gene flow to operate, preventing the species from changing into a new species.

Although gradual evolution and punctuated equilibrium are alternative explanations for the evolution of new species, one does not necessarily exclude the other. The questions now being debated among many biologists are whether one phenomenon occurs more frequently than the other, and which one most likely occurred in the evolution of a particular group.

Evolution does not necessarily progress at a constant pace but may take place in spurts, separated by periods where little change occurs. (See CTQ #7.)

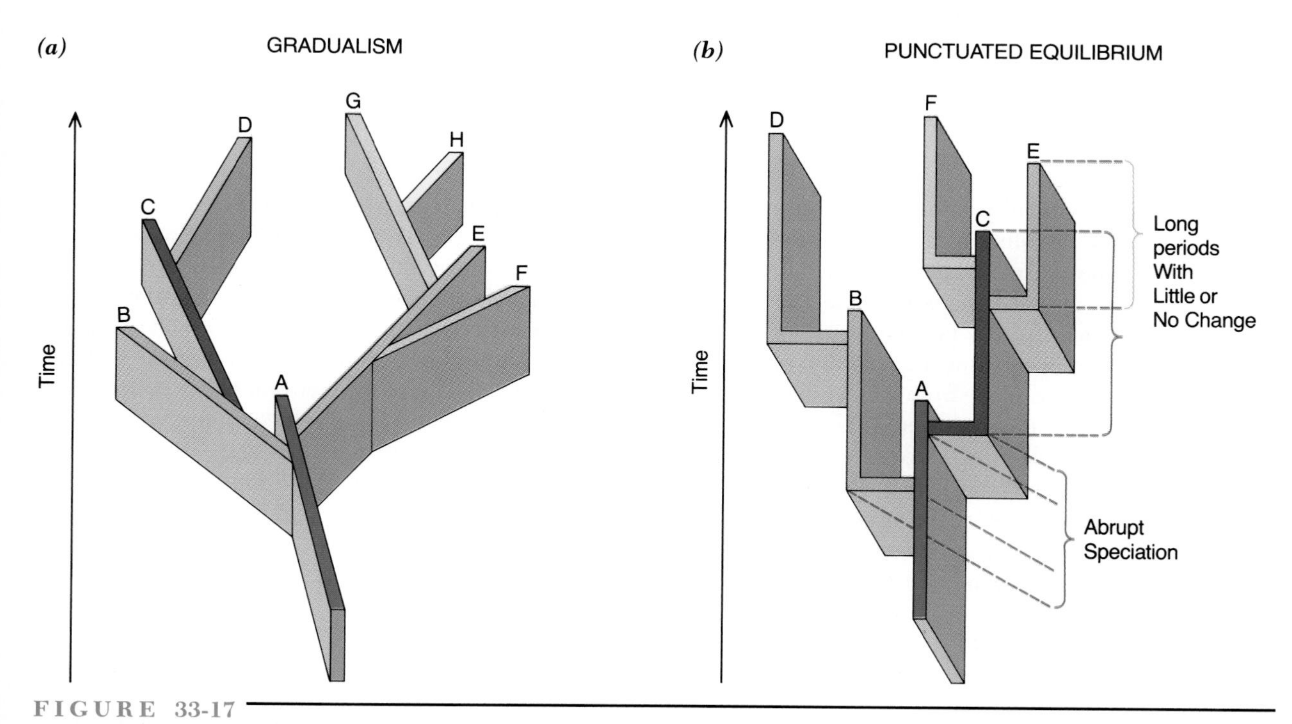

FIGURE 33-17

Gradualism versus punctuated equilibrium. ***(a)*** During gradualism, species arise through gradual, steady accumulation of changes. ***(b)*** During punctuated equilibrium, species arise as a result of the rapid accumulation of changes. Once formed, the species remains relatively unchanged for long periods of time.

REEXAMINING THE THEMES

Relationship between Form and Function

Form, function, and natural selection are inseparably linked to one another. Genes code for proteins that determine the form of most structures. Normally, when mutations arise that disrupt the structure of an essential protein, the organism fails to survive and reproduce. As a result, natural selection has eliminated the organism's genes from the species' gene pool. This is not always the case, however, as exemplified by the sickle cell allele (S), in which the alteration of red blood cell structure makes a person carrying one S allele resistant to the ravages of malaria. Consequently, the S allele has been selected for rather than being selected against.

Unity within Diversity

Evolution accounts for both the unity and diversity of life on earth. Genetic changes in populations leads to speciation, which increases the variety of species on earth. Over longer periods of time, evolutionary lineages tend to diverge from one another, leading to greater diversity. At the same time, since all organisms are related by common descent, they share many of the same basic properties of life, including a common system of storage and utilization of genetic information and a common set of metabolic pathways. Unity and diversity are also seen in the characteristics of a single species. Although all members of the species are unified by their capability for interbreeding, the individuals may exhibit a diverse morphology. This is illustrated by the variations in coloration among sea stars and female swallowtail butterflies; the body shape among dogs; and facial appearances, pigmentation, and blood type among humans.

Evolution and Adaptation

Changes in the characteristics of organisms result from changes in the frequency of particular alleles in a population. Changes in allele frequency result from several agents: mutation, genetic drift, gene flow, natural selection, and nonrandom mating. Of these various agents, only natural selection leads to the formation of organisms that are better adapted to their environment and, thus, are more likely to survive to reproductive age.

SYNOPSIS

Evolution is the process whereby species become modified over generations. Evolution occurs when the frequency of alleles in the gene pool of a population changes from one generation to the next. Changes in allele frequency can result from five identifiable factors: mutation (the introduction of new alleles); gene flow (the addition or removal of alleles when individuals move from one population to another); genetic drift (alterations in allele frequency due to chance); natural selection (increased reproduction of individuals with phenotypes that make them better suited to survive and reproduce in a particular environment); and nonrandom mating (increased reproduction of individuals with phenotypes that make them more likely to be selected as mates).

Of the factors listed above, mutation and gene flow introduce new genetic material into a population, while genetic drift, natural selection, and nonrandom mating determine which alleles will be passed on to the next generation. Genetic drift is particularly important in small populations, where chance events can have a major impact on a population's gene pool. Consequently, genetic drift becomes most important when a population shrinks during a bottleneck or when a small group of individuals break away from the main population and colonize a new habitat. Natural selection is the only factor that can cause a species to adapt to its environment. Natural selection is particularly important when environments change, allowing those individuals that possess favorable phenotypes to survive and reproduce, thereby passing their alleles on to the next generation. Natural selection can have a stabilizing, directional, or disruptive effect on the gene pool of a population.

The diversity of life on earth has arisen through repeated speciation events. For speciation to occur, a population must split into two or more separate populations that can no longer interbreed. Reproductive isolation usually occurs as a result of the formation of a geographic barrier. Following reproductive isolation, the separate populations tend to diverge from one another, until they are no longer able to interbreed, and they become different species.

Several identifiable patterns of evolution can be discerned. The most common pattern, divergent evolution, occurs when one species gives rise to two or more species that become increasingly different from one another. When divergent evolution occurs in a new area with diverse environments and an absence of competitors, adaptive radiation may occur, whereby new species form and diverge, producing a variety of related species that are adapted to different habitats. In contrast, when unrelated species colonize similar habitats, they may acquire similar adaptations that cause them to resemble one another. Changes in one species can influence the course of evolution of another species. Extinction is the ultimate fate of most, if not all, species. Extinction occurs when a species lacks the genetic variability needed to adapt to a changing environment ("bad genes") or when a sudden catastrophe occurs that essentially eliminates all life in a particular habitat ("bad luck").

The pace of evolution need not be constant. Evolution within a lineage of organisms may occur gradually in small, adaptive steps, or it may occur in spurts, in which species form and remain unchanged for long periods, followed by a period of rapid change.

Key Terms

evolution (p. 717)
species (p. 717)
population (p. 717)
gene pool (p. 717)
allele frequency (p. 717)
Hardy-Weinberg law (p. 718)
genetic equilibrium (p. 718)
gene flow (p. 719)
genetic drift (p. 720)
bottleneck (p. 720)
founder effect (p. 721)
natural selection (p. 721)
stabilizing selection (p. 724)
directional selection (p.725)
disruptive selection (p. 726)
dimorphism (p. 727)
polymorphism (p. 727)
nonrandom mating (p. 727)
sexual selection (p. 727)
microevolution (p. 731)
macroevolution (p. 731)
speciation (p. 731)
biological species concept (p. 732)
isolating mechanism (p. 732)
prezygotic isolating mechanism (p. 732)
postzygotic isolating mechanism (p. 732)
phyletic speciation (p. 732)
allopatric speciation (p. 732)
parapatric speciation (p. 732)
sympatric speciation (p.733)
hybridization (p. 733)
divergent evolution (p. 734)
adaptive radiation (p. 734)
convergent evolution (p. 735)
parallel evolution (p. 735)
coevolution (p. 735)
extinction (p. 736)
mass extinction (p. 736)
gradualism (p. 737)
punctuated equilibrium (p. 737)

Review Questions

1. Match the term with its definition.

____1. allopatric speciation
____2. disruptive selection
____3. sympatric speciation
____4. gene pool
____5. gene flow
____6. genetic drift
____7. extinction
____8. convergent evolution

a. formation of a species by geographic isolation
b. formation of a species by ecological isolation
c. result of emigration and immigration
d. change in gene frequencies due to chance
e. organisms resemble each other because of similar adaptive pressures, not common ancestry
f. extreme phenotypes in a species leave more offspring than do average phenotypes
g. the death of every member of a species
h. all of the alleles in all of the members of a species.

2. Of all the factors that cause allele frequency to change over time, why is natural selection the only one that leads to increased adaptation to the environment?
3. Why must gene flow stop before speciation can occur?
4. Why do genetic drift and gene flow have a greater impact in changing the gene frequencies of a small population than of a large one?
5. Did speciation occur in the peppered moth populations of England? Under what conditions might speciation occur in the moth?

Critical Thinking Questions

1. Of the five factors that can affect allele frequencies in a population, which could have been important for insects in developing resistance to pesticides? Which could have been unimportant? Why?
2. Populations that are not changing must meet the five conditions identified by Hardy and Weinberg. Consider each of these conditions for the case of sickle cell anemia among African Americans. Do any of the five apply? If so, which one(s)? Based on your analysis, would you predict that change in the frequency of the sickle cell is or is not occurring among African Americans?
3. New reproductive technologies, such as improved artificial insemination, have revolutionized the management of domestic animals. For example, many of the dairy cows in the United States have the same father or grandfather. What is the evolutionary disadvantage of having such a small number of fathers for the population?
4. Two very similar squirrels are found on the north and the south rims of the Grand Canyon. The Kaibab squirrel of the north rim is distinctly darker than the Abert squirrel of the south rim, however. Interbreeding occurs rarely, if ever, in nature, but could occur between members of the two squirrel species, producing fertile offspring. Would you consider these squirrels members of the same or different species? Why, or why not?
5. Match the examples below with the following patterns of evolution: divergent evolution; adaptive evolution; phyletic evolution; convergent evolution; coevolution; parallel evolution. For each one, explain why the example fits the pattern. (1) bears and pandas (2) wolves and foxes (3) the yucca plant and the yucca moth (4) *Homo erectus* and *Homo sapiens* (5) ostrich and emu (6) 16 species of Hawaiian honeycreepers evolved from a common ancestor, each with a different niche.
6. The graph below shows the numbers of extinct species and subspecies of vertebrates from 1760 through 1979. What factors have caused the tremendous increase in extinctions? How do these extinctions differ from the usual extinctions that are a natural part of evolution?

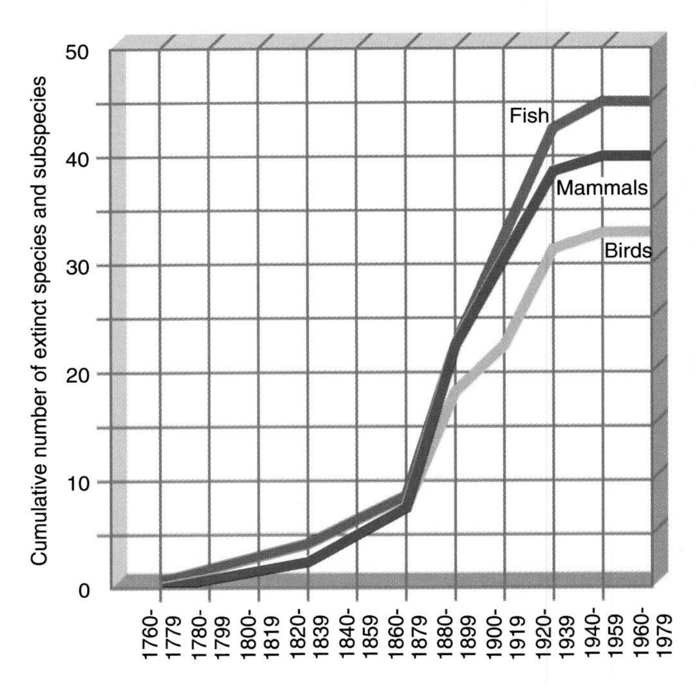

7. How would a scientist who supports gradualism explain gaps in the fossil record? How would a scientist who supports punctuated equilibrium explain the gaps? What sort of evidence would you need to convince you to accept gradualism rather than punctuated equilibrium, or vice versa?

Additional Readings

Avers, C. J. 1989. *Process and pattern in evolution.* New York: Oxford University Press. (Intermediate)

Carson, R. 1962. *Silent spring.* Boston: Houghton Mifflin. (Introductory)

Cook, L. M., G. S. Mani, and M. E. Varley. 1986. Postindustrial melanism in the peppered moth. *Science* 231:611–613. (Advanced)

Dodson, E. O., and P. Dodson. 1985. *Evolution: Process and product,* 3rd ed. Belmont, CA: Wadsworth. (Advanced)

Futuyma, D. J. 1986. *Evolutionary biology,* 2nd ed. Sunderland, MA: Sinauer. (Advanced)

Gould, S. J. 1992. What is a species? *Discover* Dec:40–44. (Intermediate)

Marco, G. J., R. M. Hollingworth, and W. Durham. 1987. *Silent spring revisited.* Washington, D. C.: American Chemical Society. (Introductory)

O'Brien, S. J., D. E. Wildt, and M. Bush. 1986. The cheetah in genetic peril. *Sci. Amer.* 254:84–95. (Intermediate)

Shell, E. R. 1993. Waves of creation. *Discover* May:54–61. (Intermediate)

Wills, C. 1989. *The wisdom of the genes: New pathways in evolution.* New York: Basic Books. (Intermediate)

CHAPTER

34

Evidence for Evolution

STEPS
TO
DISCOVERY
An Early Portrait of the Human Family

STEPS TO DISCOVERY

An Early Portrait of the Human Family

Scientists use the term **hominid** to refer to humans and the various groups of extinct, erect-walking primates that were either our direct ancestors or their relatives. The first fossil remains of a hominid were unearthed in 1856 in caves of the Neander Valley in Germany. After much debate, the remains were dismissed as the bones of a deformed Russian soldier who had died in an earlier war with France.

After the discovery of similar bones in other locations around Europe, it became apparent that the earth had been inhabited at one time by "people" that resembled humans but possessed noticeable differences. They were called Neanderthals, after the site where they were first discovered. Their skulls had a shape different from that of modern humans, with heavy, bony ridges over the eyes, and their bones were much thicker, with indications of larger attached muscles. The Neanderthals were depicted in the popular press as brutish-looking, grunting, stooped-over cavemen. In reality, if you were to see one of these beings walking down the street in jeans and a T-shirt, you probably wouldn't turn and take notice. Neanderthals lived between 35,000 and 135,000 years ago.

The first evidence of a hominid that would cause you to take notice if you saw one walking down the street was discovered in 1891 by Eugene Dubois, a doctor in the Dutch army stationed in the Dutch East Indies. Soon after arriving on the island of Java, Dubois found the remains of extinct mammals. One day, he found a back (molar) tooth that he thought must have belonged to an ape. A meter away, he discovered a skull that possessed characteristics of both human and ape anatomy. The next year, approximately 15 meters from where he had found the skull, Dubois unearthed a thigh bone (femur) that was very similar to that of

Archaeologists search fossil beds in Africa for remains of early humans illustrated in the time bubble.

a modern human. Most importantly, the shape of the femur indicated that the owner had walked erect. Dubois concluded that he had found the "missing link." He packed the pieces of his "Java Man" into a box and returned triumphantly to Europe.

Most of the scientific world greeted Dubois' claims with skepticism. Sir Arthur Keith, one of the most prominent paleontologists of the time, had a different opinion. After examining the fossils closely, Keith concluded that, even though the size of the Java Man's braincase (the part of the skull that covers the brain) was not much larger than that of an ape, the skull showed definite human features. Keith was so convinced of the similarities that he recommended the Java Man be placed into the same genus as are modern humans. Eventually, Dubois' find became designated *Homo erectus* (see Figure 34-8). However, Dubois never accepted Keith's view that Java Man should be classified as *Homo* (which, in essence, describes it as a human). In response, Dubois buried the bones of his missing link under the floorboards of his dining room, where they remained for the next 30 years.

Over the next 30 to 40 years, a number of other fossils were found that were similar to that of Java Man and were also assigned to the species *H. erectus.* The most important find was Peking Man, discovered in a cave near Peking, China. Like Java Man, Peking Man had a small, apelike braincase; thick, heavy bones; a prominent, bony ridge above the eyes; and a humanlike lower jaw with humanlike teeth. Most importantly, it was demonstrated that Peking Man had walked with an erect posture, used stone tools, and cooked his dinner over a fire. Both Java Man and Peking Man lived about half a million years ago.

Two fossil finds did not fit the profile of *H. erectus,* however. One was a remarkably complete skull that was discovered by an amateur fossil hunter in 1912 near the town of Piltdown in England. The skull of this so-called Piltdown Man had a large braincase (as large as that of a modern human) and an apelike jaw, characteristics in direct contrast to those of Java Man and Peking Man. Piltdown Man presented a serious problem for interpreting the path of human evolution. Some paleontologists dismissed Piltdown Man as an anomaly. Others, including Sir Arthur Keith and the British anthropological establishment, embraced Piltdown Man as an important fossil and suggested that the development of a large brain may have been one of the earliest characteristics to appear along the path of human evolution.

The other perplexing fossil discovery was made in 1924 by Raymond Dart, an Australian on the faculty of a medical school in Johannesburg, South Africa. Dart heard that fossils were being uncovered at a limestone quarry in an area of South Africa called Taung. He asked the owner of the quarry if he might see some of the fossils; two large boxes were shipped to his house. As he was pouring through the contents, Dart spotted a dome-shaped piece of stone. As a neuroanatomist, he immediately recognized the stone as the cast of a brain, complete with indications of convolutions and blood vessels. Sand and lime-containing water had seeped into the skull of an ancient inhabitant of the quarry and hardened, forming a cast of what had once been the creature's brain. Although the brain was the size and form of an ape's, it revealed distinct humanlike characteristics. Dart began searching for the skull that had recently surrounded the cast, believing that it must have been blasted away during the mining operation.

Among the contents of the box, Dart found the remains of the lower jaw and skull, the front of which was covered by an encrusted material, making it impossible to see the face. For the next couple of months, Dart carefully picked away at the crust and slowly revealed an astonishing visage; it was the face of a young "ape," with teeth that showed striking human characteristics (see Figures 34-8 and 34-9). The cranium was slightly larger than that of an ape, and the opening in the skull that allowed entry of the spinal cord was in a position different from that of an ape, suggesting that the individual had walked erect. Based on the other fossils in the box, Dart concluded that the skull was about 1 million years old. He named the creature *Australopithecus africanus* (*Australo* = southern, *pithecus* = ape), but it became known as the Taung Child.

Without delay, Dart wrote up a paper on his skull and sent it to the prestigious British journal *Nature,* in which it was published. Once again the scientific world was very skeptical. Even Keith, who was a friend of Dart's, clung to the notion (based primarily on Piltdown Man) that enlargement of the brain was one of the first steps in hominid evolution and declared that the Taung Child was not a hominid but an extinct ape.

In 1931, Dart traveled to London to attend an anthropological meeting in hopes of convincing his colleagues of his claim. Dart's talk followed a dazzling presentation of the findings that were emerging from China concerning Peking Man. In addition, Dart was a poor speaker, whose evidence was limited to a single skull; he failed to make much of an impact. Discouraged, he went off to dinner with friends while his wife brought the Taung Child back to the hotel. As if the day hadn't gone badly enough, his wife left the infamous skull (wrapped in cloth) in the back seat of the cab, where it traveled around London most of the night. The cab driver finally saw the package and handed it over to the police. Fortunately, Dart was able to recover his package before the police had time to wonder what type of skullduggery they had on their hands.

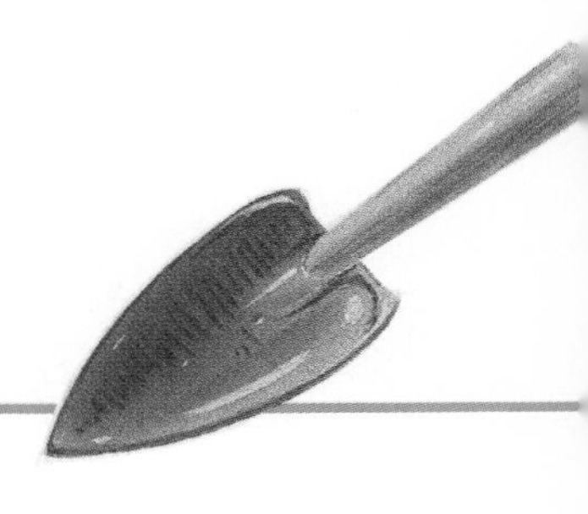

The "theory of evolution" is no less a fact of life to biologists than the "atomic theory" is to chemists or the "theory of gravitation" is to physicists. For a theory of such importance to have gained such widespread acceptance, it must be backed by a tremendous body of evidence. We will begin by sampling a small portion of this evidence, taken from a wide variety of different fields. The entire matter can be summed up in a single sentence written by the biologist Theodosius Dobzhansky: "Nothing in biology makes sense except in the light of evolution."

▼ ▼ ▼

DETERMINING EVOLUTIONARY RELATIONSHIPS

Given that closely related species share a common ancestor and often resemble one another, it might seem that the best way to uncover evolutionary relationships would be to make comparisons of overall similarity between organisms. In other words, out of a group of species, if two are most similar, can we reasonably hypothesize that they are the closest relatives? Surprisingly, this is not always the case (Figure 34-1). Overall similarity may be misleading because there are actually two reasons why organisms may have similar characteristics, only one of which is due to evolutionary relatedness.

HOMOLOGOUS VERSUS ANALOGOUS FEATURES

Two species that share a similar characteristic they inherited from a common ancestor are said to share a **homologous feature,** or **homology.** The even-toed foot of deer, camels, cattle, pigs, and hippopotamuses, for example, is a homologous feature because all of these animals inherited the characteristic from a common extinct ancestor (Figure 34-2). When *unrelated* species evolve a similar mode of existence, however, their body parts may take on similar functions and end up resembling one another due to convergent evolution (page 735). This type of shared characteristic is called an **analogous feature,** or **homoplasy.** The paddlelike front limbs and streamlined bodies of many aquatic animals (see Figure 33–15) are examples of analogous features.

▮▮▶ Homologous similarity is the only evidence that proves that two species are evolutionarily related. But how do biologists tell whether a similarity is homologous or homoplasious? Years of experimentation and observation have resulted in a set of criteria that are used to identify homologies. These criteria include: (1) similar in detail, (2) similar position in relation to neighboring structures or organs, (3) similarity in embryonic development, and (4) agreement with other characters (related animals usually share more than one homology).

These criteria of homology can be illustrated by examining a variety of mammalian forelimbs (Figure 34-3). At first glance, the wing of a bat, the leg of a cat, the flipper of a whale, the arm of a human, and the leg of a horse may not seem very similar, but they are actually homologous. All of these limbs contain the same type of bones (similar in detail); the forelimb always attaches to the shoulder girdle (similar position in relation to neighboring structures); the forelimb develops from the same tissues in each of the embryos (similar embryonic development); and, in addition to the forelimb, all these animals have hair and mammary glands (other shared homologies).

Evolutionary relationships cannot be reconstructed just by grouping species together by their number of shared homologies, however. For example, the hand of the first vertebrates to live on land had five digits (fingers). Many terrestrial vertebrates (such as humans, turtles, lizards, and frogs) also have five digits, which they inherited from this common ancestor. This feature is then homologous in all of these species. In contrast, horses, zebras, and donkeys have only a single digit with a hoof. But, clearly, humans are more

(a)

(b)

FIGURE 34-1
Deceptive similarities. These two "palms" could easily be mistaken for closely related plants based on their similar appearances, but the cycad *(a)* is not a palm, or even a flowering plant. It is no more related to the true palm *(b)* than is a pine tree to a rose.

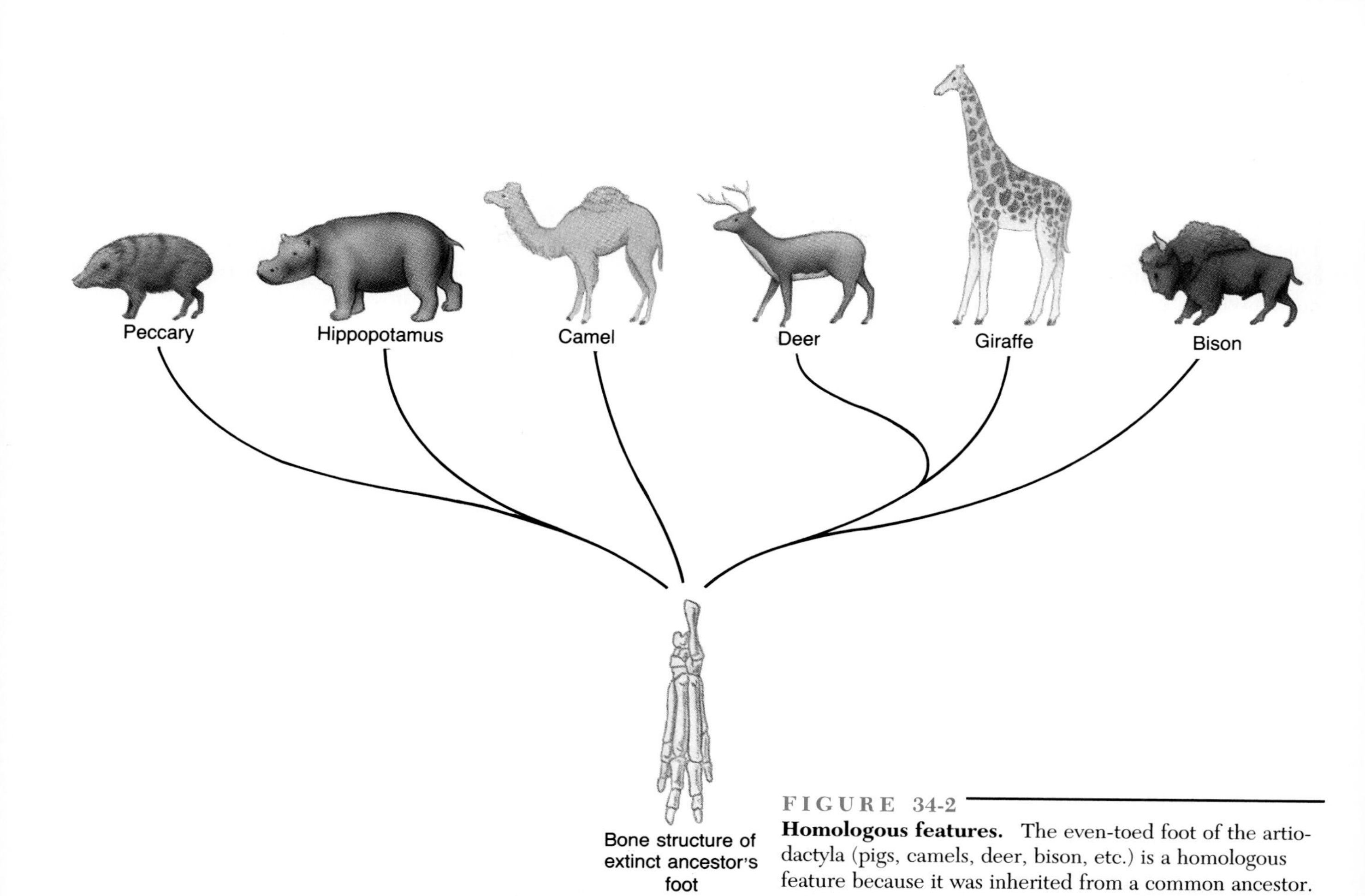

FIGURE 34-2
Homologous features. The even-toed foot of the artiodactyla (pigs, camels, deer, bison, etc.) is a homologous feature because it was inherited from a common ancestor.

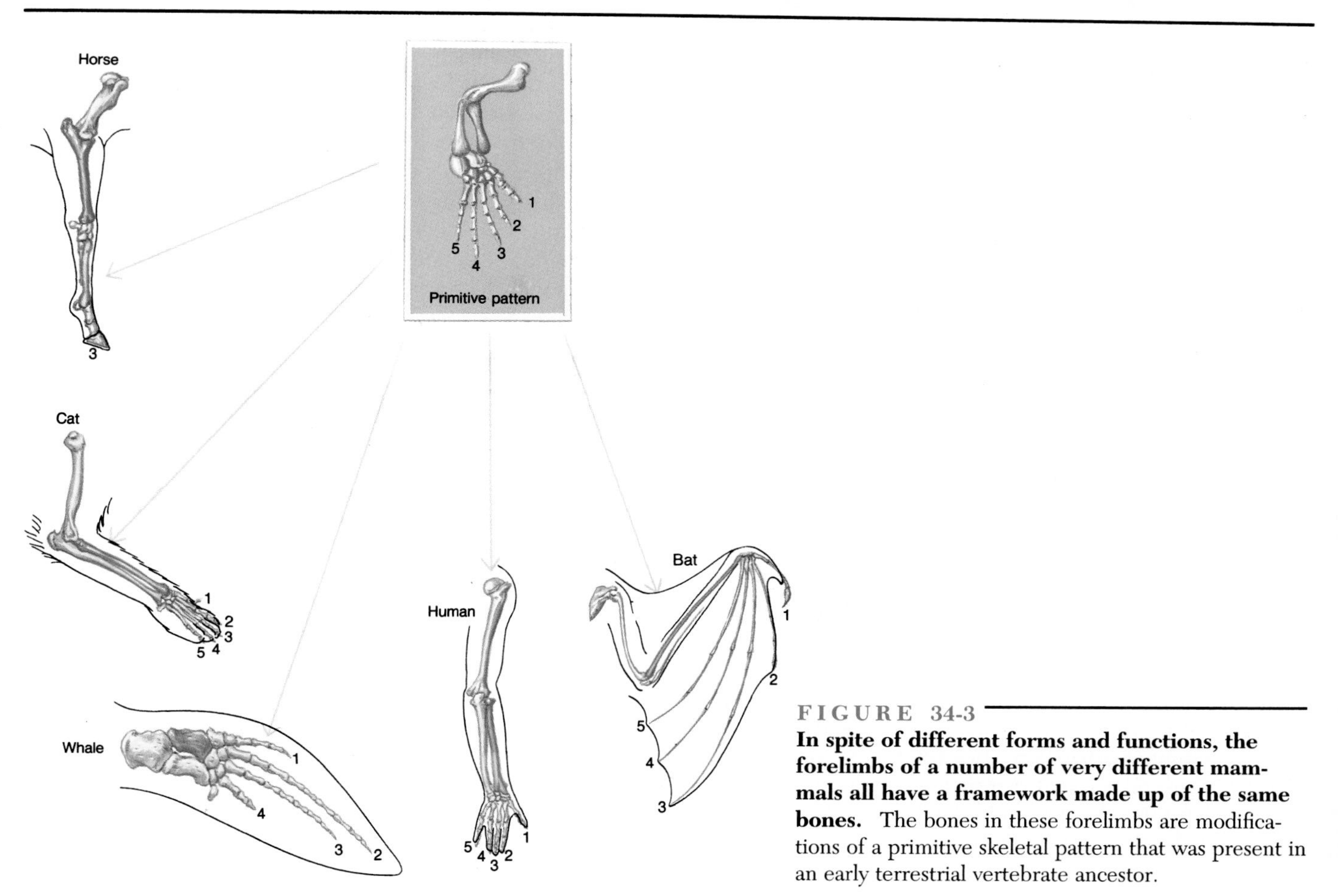

FIGURE 34-3
In spite of different forms and functions, the forelimbs of a number of very different mammals all have a framework made up of the same bones. The bones in these forelimbs are modifications of a primitive skeletal pattern that was present in an early terrestrial vertebrate ancestor.

closely related to horses than they are to lizards! The key point is that the five-digit condition is the *primitive* (original) pattern for the number of digits (shown in Figure 34-2). This primitive feature has been retained in the line of ancestors leading from early amphibians to humans, but it has been modified and reduced to just one digit in the common ancestor of horses, donkeys, and zebras. While the *derived* (modified) trait tells us that horses, zebras, and donkeys share a very recent common ancestor, the primitive form (five digits) tells us only that species are at least distantly related.

Let us now turn to the types of data that biologists use to distinguish between homologies and homoplasies and between primitive homologies, which may be present in rather distantly related species, and shared derived homologies, which indicate closer relationships.

Appearances can be deceiving. Biologists must apply a number of criteria to determine if similarities in structural features are the products of an evolutionary relationship or the consequence of similar selective pressures exerted on unrelated organisms. (See CTQ #2.)

EVIDENCE FOR EVOLUTION

According to the theory of evolution, organisms living today have arisen from earlier types of organisms by a process of genetic change that has occurred over a period of several billion years. The fact that all organisms have arisen from a common ancestor explains why they have the same mechanism for the storage and utilization of genetic information, many of the same types of cellular organelles, and similar types of enzymes and metabolic pathways. At the same time, evolution also explains how a single species can give rise to numerous other species, leading to great biological diversity. Evidence supporting the theory of evolution has accumulated from a variety of different biological disciplines, including comparative anatomy, paleontology, comparative embryology, biochemistry and molecular biology, and biogeography.

FOSSIL RECORDS

A fossil is any trace of life from the past. Many different types of fossils exist, ranging from preserved "footprints" of animals that walked along a trail (see Figure 34-12) to complete remains, such as frozen mammoths or entombed insects (Figure 34-4*a*), to actual hard parts (teeth and bones), to pieces of petrified trees (Figure 34-4*b*) and even preserved excrement (*corpolites*). Fossils of body parts may be formed in several ways: Organisms may be buried in sediments, where they harden and mineralize; trapped in tree sap, which hardens into amber; covered in tar or other natural preservatives, such as the liquid found in peat bogs; or frozen in arctic regions or at high altitudes.

The **fossil record** consists of an entire collection of such remains from which paleontologists attempt to reconstruct the biology of the organisms whose remains were left behind. In addition to providing a glimpse of the kinds of organisms that lived in the past, the fossil record provides data evolutionary biologists can use to describe the pathways by which various groups may have evolved. For example, without fossils, we would not realize the close relationship between dinosaurs and birds and the fact that birds are more closely related to reptiles than they are to mammals.

Fossil Dating

For many years, scientists had no way of directly determining the age of a fossil. Instead, scientists would establish the sequence of fossils embedded in the layers of rock from top to bottom and then apply the *law of stratigraphy*. According to this law, in beds of rocks that have not been tilted or folded, the oldest rocks are always on the bottom, and the youngest rocks are always on top. In some sites, rocky strata covering tens of millions of years are present in layers, containing the remains of organisms in chronological order, much like a giant filing cabinet in which the more recent documents are found in drawers situated closer to the top. For example, while remains of ancient fishes are common in rocks 400 million years old, no evidence of reptiles has ever been found in such sediments. Similarly, remnants of reptiles may be unearthed in 285-million-year-old rocks, but we never find evidence there of a bird or mammal. In H. G. Wells's words: "The order of descent is always observed." In the uppermost sediments on earth, biologists may find the remains of animals that have only recently become extinct, including creatures with apelike faces that walked on two feet and crafted primitive stone tools.

While stratigraphy still plays an important role in determining which types of organisms coexisted, the absolute age of fossils can now be determined with a high degree of accuracy using radiodating techniques. As we discussed on page 54, these techniques depend on the existence of naturally occurring radioisotopes that disintegrate into other elements at a predictable rate. Using radioisotopes, scientists have shown that the oldest rocks on earth are over 4 billion years old, much older than Charles Darwin could even have imagined when he first realized that the earth was older than that determined by strict interpretation of the Bible.

Archaeopteryx: An Example of Fossil Evidence

One of the best known fossils was discovered in 1861 in a limestone quarry in Bavaria, Germany. The skeleton of the fossil (Figure 34-5*a*) suggested that the animal had been a small bipedal dinosaur, but the fine-grained limestone slate also revealed the unmistakable imprint of wings with feathers. Of all the vertebrates, only birds possess feathers; in fact, feathers are a defining characteristic that unites all

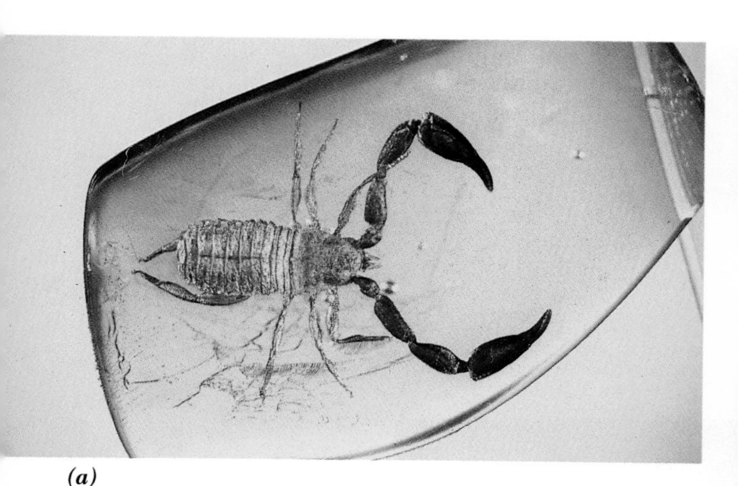

(a)

(b)

FIGURE 34-4

Types of fossils. ***(a)*** This ancient pseudoscorpion (a distant relative of spiders) was trapped in a drop of resin that became transformed into hardened amber. ***(b)*** A scene from the Petrified Forest of Arizona.

birds. This animal, which was given the name *Archaeopteryx lithographica* (*archaeo* = old, *pteryx* = wing), was determined to have been a bird that lived 150 million years ago. Yet, unlike all modern birds, *Archaeopteryx* had teeth, a long tail containing over 20 vertebrae, free-floating ribs, and wings containing movable fingers with claws, all characteristics of the small, carnivorous reptiles called *theropods.* Therefore, *Archaeopteryx* provides one of the many pieces of fossil evidence of an evolutionary pathway leading from reptiles to birds.

Although the skeleton of *Archaeopteryx* has a wishbone (two collarbones fused together), which is typical of birds, it lacks the broad, bony breastbone to which the large flight muscles of modern birds are attached. This and other skeletal characteristics reveal a great deal about the lifestyle of this ancient vertebrate. The lack of a breastbone suggests that *Archaeopteryx* was not a strong flyer and may have been primarily a glider. The claws on the toes suggest that the animal could perch on a limb, while the claws on the "fingers" suggest that it may have been able to climb up the trunks of trees (Figure 34-5*b*). The structure of the pelvis and hind legs suggests that *Archaeopteryx* was capable of running over the ground, its long tail acting as a counterweight to maintain its balance. The teeth of this pigeon-sized animal would have been suitable for the capture of insects or other small prey.

Cautions Regarding the Fossil Record

The fossil record provides an imperfect view of ancient life because not all organisms and environments are represented equally. For example, fossils of organisms with hard parts, such as shells or skeletons, or the woody parts of plants are more common than are fossils of soft-bodied organisms. Thus, arthropods, which possess hardened exoskeletons, are much more abundant in the fossil record than are jellyfish. Similarly, the likelihood of fossilization is much greater when an organism lives in an environment in which bodies or impressions of bodies can be covered quickly with sediments (such as in shallow seas and river beds) than in other environments.

Furthermore, there is only a remote chance that a fossil will ever be found because only a small proportion of the fossil-bearing rocks are accessible to us. Most fossil-bearing rocks have been eroded away, buried deep beneath the continents or the ocean floor, or broken and destroyed by the earth's movement. Even when a fossil is found, it is

(a)

(b)

FIGURE 34-5

Archaeopteryx, a bird that lived 150 million years ago, has many features in common with small, bipedal dinosaurs—features, such as teeth and a long tail, that were eliminated during the evolution of modern birds. ***(a)*** Photograph of the fossil imprint of *Archaeopteryx* in limestone slate. ***(b)*** An artist's rendition of the long-extinct bird as it may have appeared in life.

often only a fragment of a bone or shell, inviting speculation regarding its significance. In general, the further back in time we go, the less complete the fossil record.

Interpreting the fossil record is often difficult, and conclusions are open to speculation. Although some organisms in the fossil record are indeed ancestors of living organisms, an older fossil is not necessarily the ancestor of a younger fossil, and we can't assume an evolutionary lineage or rate of speciation from a fossil sequence. This point is illustrated by the story of human evolution. Most of the fossils that have been found are probably an *offshoot* from the direct line of descent of modern human beings. While there is only one erect-walking, tool-making, word-speaking species of hominid living today, 2 million or 3 million years ago, there may have been quite a number of such species. It is impossible to establish a straight, ladderlike lineage leading from this "bush" of species to a single living representative.

THE ANATOMY OF LIVING ORGANISMS

III▶ Comparing the structure of the parts of the bodies of different organisms is probably the most commonly used evidence of evolution. In order to gather comparative evidence for evolution, biologists study external characteristics, examine bones and teeth, dissect organ systems, study sections of tissue under the light microscope, and peer at the finer details of cells and tissues under the electron microscope. Along the way, much is added to our knowledge about the basic biology of different organisms.

Vestigial Structures

The underdeveloped pelvis and leg bones in snakes, the diminished toe bones in horses, and the appendix in humans are all structures that have little or no function in these organisms (Figure 34-6). Each of these **vestigial structures,** as they are called, bears an unmistakable stamp that shows its relationship to a more fully developed, functional structure present in other animals. For example, your appendix is a dwarfed version of the cecum, a part of the digestive tract of many mammals, where food is stored and digested by microorganisms. Similarly, your tailbone is a remnant of the tail present in many of your relatives.

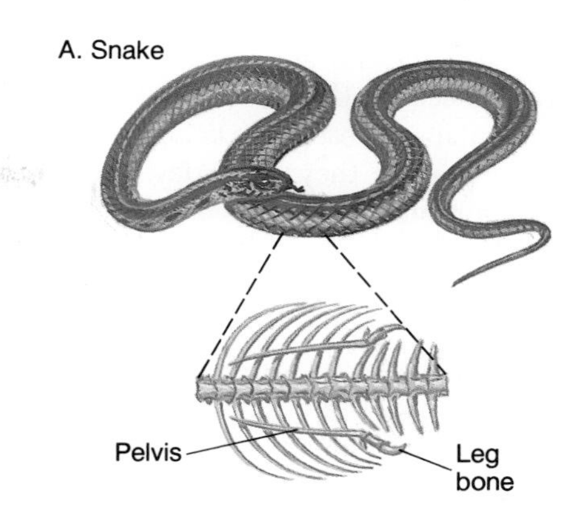

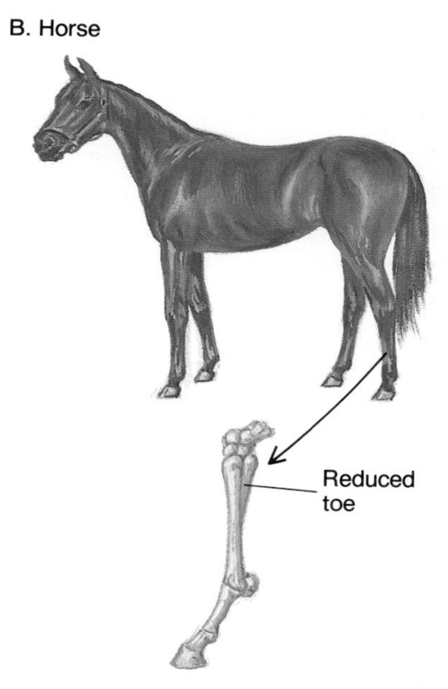

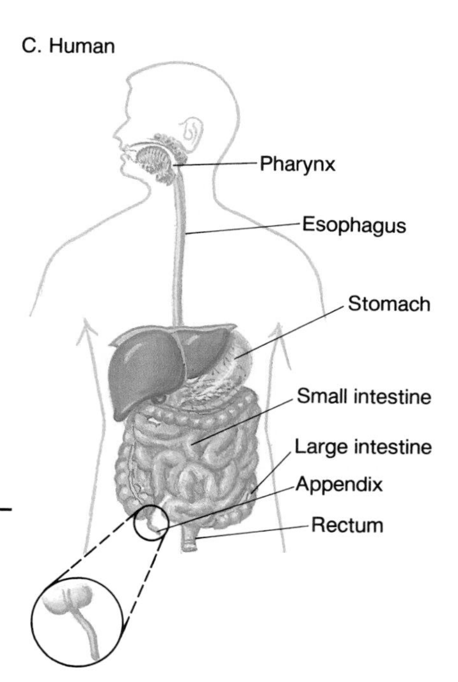

FIGURE 34-6

Vestigial organs: A legacy of ancestors. ***(a)*** Even though they are legless, snakes retain degenerated leg and pelvic bones inherited from four-legged ancestors. ***(b)*** Horses still possess degenerated toe bones left over from their three- and four-toed ancestors. ***(c)*** The human appendix is a degenerated cecum, a chamber used by our vegetarian ancestors for housing cellulose-digesting microorganisms.

Both the plant and animal kingdoms are filled with examples of structures that are in the process of disappearing but still remain as vestiges in living organisms. If species had been created as they are today, there would be no explanation for such structures, but natural selection accounts for the elimination of structures that are no longer needed by an organism exploiting a different way of life.

COMPARATIVE EMBRYOLOGY

Adult fishes, salamanders, turtles, birds, and humans bear virtually no resemblance to one another. Yet these animals are virtually indistinguishable as embryos (see Figure 32-18). Why should animals that have markedly different adult forms and functions develop from such similar embryos? The best scientific explanation is that far back in vertebrate history, fishes, salamanders, turtles, birds, and humans all had a common ancestor, probably some type of primitive fish, that developed from a similar type of embryo. As the various types of vertebrates evolved, they each retained this basic vertebrate embryo as part of their life cycle, even though its parts gave rise to different adult organs.

This example illustrates how developmental evidence can be particularly useful in uncovering evolutionary relationships among diverse adult forms. For example, embryologic similarities between widely separated animal groups (such as snails and flatworms) have been used to reconstruct lines of evolutionary descent within the animal kingdom.

There is another way to illustrate the constraints of embryonic development. Recently, Harvard paleontologist and essayist Stephen J. Gould noted that the question "Why do men have nipples?" heads the list of inquiries from his readers. Gould argues that nipples in men are not adaptive, nor did they evolve from structures that were adaptive in an ancestor. Rather, Gould concludes that nipples in men are the result of a constraint imposed by embryonic development. Male and female mammals of a species pass through identical stages as early embryos; it is not until the secretion of sex hormones that male and female sexual development diverge. Nipples are present in human embryos *prior* to the time of sexual differentiation. Later sexual maturation leads to changes in the female breast and nipple that allow these structures to nourish a newborn infant. In contrast, the male nipple remains as the vestige of an embryonic structure that is simply carried along "for the ride" in the adult. According to Gould, "Male mammals have nipples because females need them . . ."

BIOCHEMISTRY AND MOLECULAR BIOLOGY

Since the characteristics of an organism are determined by its genetic content, changes in organisms over the course of evolution are reflected in changes in the nucleotide sequence of DNA (genes) and the amino acid sequences of proteins (gene products). For the most part, the longer the period since two species have diverged from a common ancestor, the greater the number of substitutions that are found in corresponding genes and proteins between the two species. Common ancestry can now be demonstrated just as forcefully by homologous molecular information as by homologous anatomic structures.

We saw on page 329 how this type of data has been used to determine that humans are more closely related to chimpanzees than are chimpanzees to gorillas. Molecular data of this type allow biologists to determine phylogenetic relationships among organisms based on the degree of nucleotide or amino acid sequence similarity. These diagrams are generally in keeping with conclusions based on anatomic data. As illustrated in the following example, molecular data have also been used to resolve evolutionary controversies.

What Is a Bat? The Use of Molecular Data in Studying Evolution

Biologists recognize two major groups of bats. The microchiroptera include the numerous bats commonly seen swooping up insects at night as well as the more exotic frog-eating bats and vampire bats of Central and South America (Figure 34-7*a*). The megachiroptera, also called flying foxes because of their foxlike faces, are large fruit-eating bats that live in the tropics (Figure 34-7*b*). Bats are highly specialized for flight, and many of the features that might reveal their closest relatives have been modified almost beyond recognition by natural selection, resulting in a debate among morphologists and paleontologists over the evolution of bats. One group of scientists determined that the skeletal evidence pointed to a close relationship between microchiroptera and megachiroptera and that both groups were distantly related to the primates (Figure 34-7*c*). Other scientists claimed that evidence derived from studying the nervous system indicated that the megachiroptera and the primates were closest relatives (Figure 34-7*d*). There seemed to be no data to solve the debate until late 1991, at which time molecular biologists determined the sequence of the DNA that codes for ribosomal RNA of all three groups. The molecular evidence showed that the microchiroptera and the flying foxes were indeed closest relatives, as reflected in the phylogenetic scheme of Figure 34-7*c*.

BIOGEOGRAPHY: THE GEOGRAPHICAL DISTRIBUTION OF ORGANISMS

Among the evidence that convinced Darwin of the occurrence of evolution were the observations he made on the Galapagos Islands (Chapter 1). Darwin noted that species present on oceanic islands were not found anywhere else in the world. In fact, many were found only on a particular

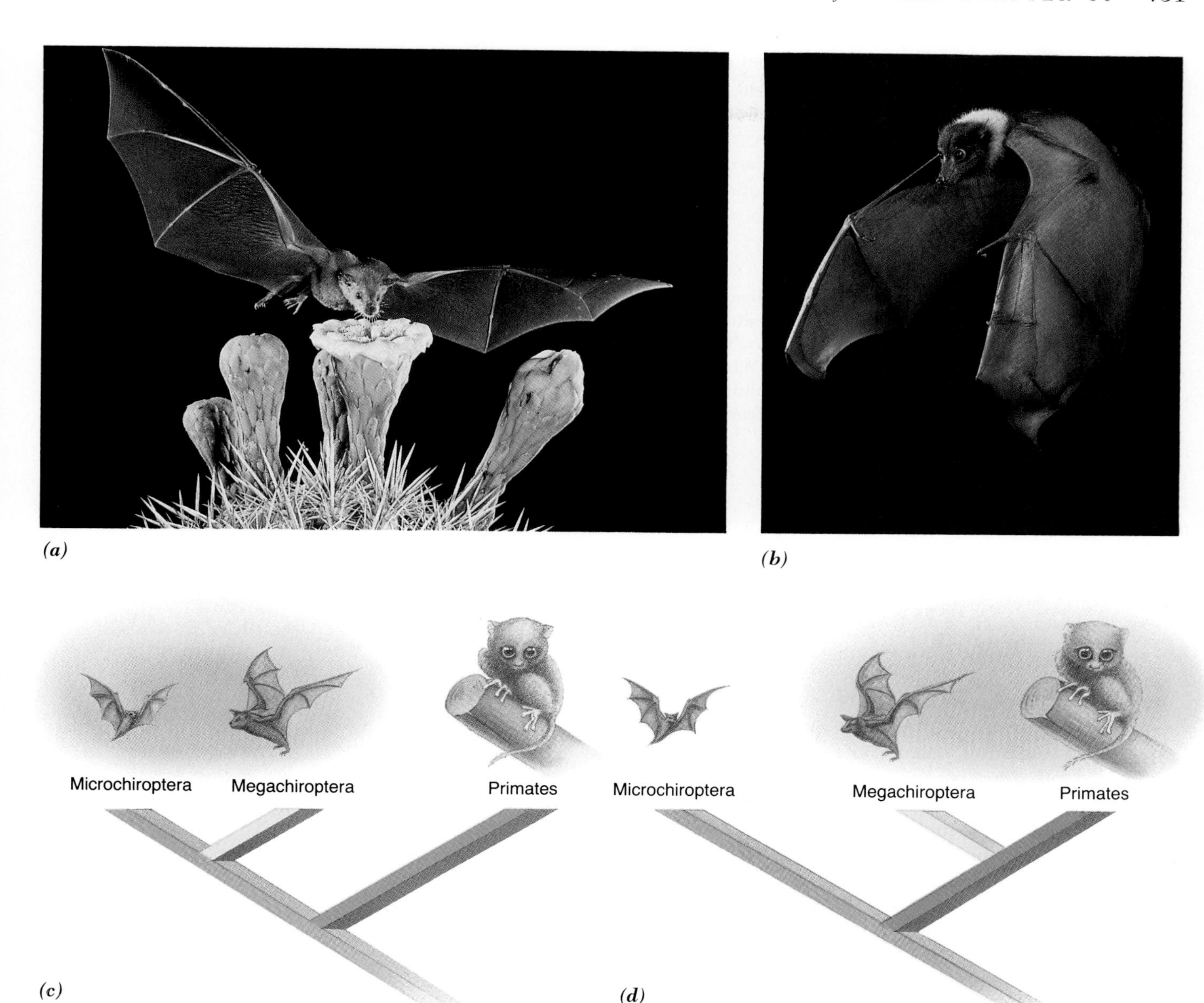

FIGURE 34-7
Evolution of bats. Evolutionary relationships of bats has been the subject of recent debate. The microchiroptera, such as the long-nosed bat ***(a)***, have been thought to be related to the megachiroptera, or flying foxes ***(b)***, but flying foxes have also been linked to primates. This conflict was resolved by molecular data that indicates that the megachiroptera are indeed more closely related to the microchiroptera (as depicted in **c**) than they are to the primates ***(d)***.

island in the chain and varied from one island to the next. Recall from Chapter 1 that Darwin found different species of finches with distinct anatomic differences living on different islands. Although the various finches possessed different-shaped beaks, which were adapted for obtaining different types of foods (see Figure 1-10), the birds were unmistakably similar in overall anatomy, both to one another and to a species found on the mainland. Darwin concluded that individuals from the mainland species had migrated to the islands, where, given the absence of competition from other birds, they had evolved into a variety of different species adapted to different local conditions and food sources. A common origin also explained why the plants and animals of the Galapagos were generally so similar to those species living on the mainland, even though the two regions had totally different climates and terrain.

These types of biogeographical observations are not restricted to the Galapagos. Plants and animals living in nearby areas typically are similar, regardless of differences in climate and terrain, because they are closely related. In fact, island and mainland species are often placed in the same genus. In contrast, plants and animals living in similar environments on different continents tend to be quite different.

The evidence for evolution comes from such diverse fields as paleontology, embryology, comparative anatomy, biochemistry, molecular biology, and biogeography. This body of evidence is both diverse and overwhelming. While biologists may argue over the mechanisms of evolution, they agree that all life descended with modification from a single common ancestor. (See CTQ #3.)

THE EVIDENCE OF HUMAN EVOLUTION: THE STORY CONTINUES

When we left the discussion of fossil hominids in the chapter-opening vignette, the story had become confused by the presence of conflicting data. On the one hand, we were confronted with *Homo erectus* and the more primitive, less well-accepted, *Australopithecus africanus* (Figure 34-8). These fossils had small, apelike brains and humanlike jaws and teeth. On the other hand, we learned of the Piltdown Man, who had a large, humanlike brain and apelike jaws and teeth.

It was not until the late 1940s and early 1950s that the matter was finally resolved. At that time, a careful analysis was conducted of the jaws of a variety of *Australopithecus* specimens that had been found in South Africa, including Dart's Taung Child (Figure 34-9) and others that had come to light over the intervening years. The analysis was performed under the leadership of Sir Wilfrid Le Gros Clark, who had become the foremost British paleontologist of the time. Le Gros Clark established 11 characteristics that clearly distinguished the teeth of humans from those of modern apes. Examination of the *Australopithecus* fossils indicated that, despite the fact that their braincase was so small and apelike, their teeth were similar to humans in every one of the 11 criteria. There was no longer any doubt that the australopithecines, as they are called, were hominids. In addition, a new radiodating technique revealed that the australopithecines were very old—up to 2 million years old. Even Keith (who was nearly 80 by this time) publicly admitted that he had been wrong and that Dart (who was still actively searching for fossils) had been right.

A second important revelation came from a more careful scrutiny of the Piltdown Man. Radiodating provided

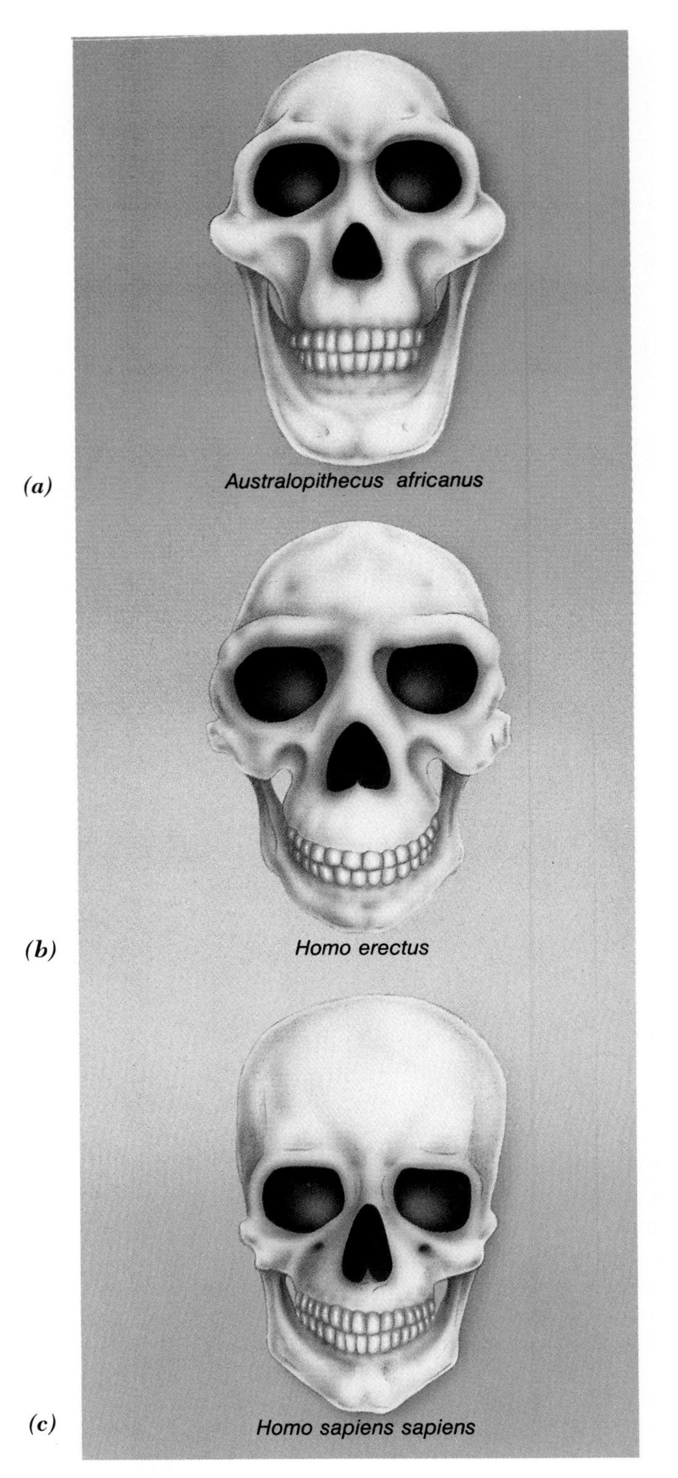

FIGURE 34-8
A comparison of the skulls of modern humans and two hominids. ***(a)*** Skull of the extinct species *Australopithecus africanus* originally represented by the Taung child and later by several other fossils found in South Africa. ***(b)*** Skull of the extinct species *Homo erectus* that includes Java Man and Peking Man. ***(c)*** Skull of a modern human, *Homo sapiens sapiens*.

FIGURE 34-9
The skull of the Taung Child in the hands of its discoverer Raymond Dart. Dart died in 1988 at the age of 95.

conclusive evidence that the Piltdown skull was actually of very *recent* vintage—a matter of a few hundred years. In other words, Piltdown Man was a hoax. Someone had taken the skull of a modern human and the lower jaw of a modern orangutan, treated them with chemicals to make them look old, filed down the ape teeth so that they resembled those of a human, broken them into fragments, and buried them alongside one another.

With Piltdown Man out of the way, it was evident that the enlargement of the brain occurred during a late stage of human evolution, not an early stage. In the past 25 years, study of Piltdown Man has shifted from the character of the bones to the perpetrator of the hoax, and fingers have been pointed at some of the leading paleontologists of the time.

HOMO HABILIS AND THE USE OF TOOLS

By the end of the 1950s, the primary scene of hominid excavation had shifted from Europe and Asia to Africa. To understand the more recent evolutionary discoveries, we need to introduce another cast of paleontologists, the most famous of which are Louis and Mary Leakey.

The Leakeys came to East Africa in the 1930s, looking for fossil hominids. They focused their attention on the now-famous Olduvai Gorge, located in Tanzania. The gorge is situated in the Serengeti Plain, on what once was a lakebed. Over time, sediments were deposited on the bottom of the lake, creating layer upon layer, the deepest sediments being the oldest. The gorge, which is about 100 meters (330 feet) deep, was created by a river that wound through the area, digging deeper and deeper into the layers of sediments. The sides of the gorge reveal the stratifications, while the bottom corresponds to the bottom of the ancient lake as it existed approximately 2 million years ago. The Leakeys were first drawn to Olduvai by the large numbers of primitive tools that were strewn over the bottom of the gorge. It was the maker of these tools for whom the Leakeys were searching.

According to the Leakeys, the use of tools is just as important (if not more important) than is brain size or tooth structure in describing a fossil hominid as a human (*Homo*), as opposed to some other genus, but not all anthropologists agree. The Leakeys were convinced that the genus *Homo* was older than was generally accepted and that the australopithecines were not our ancestors but our cousins. In other words, they believed that the members of the genus *Homo* went as far back as *Australopithecus* and that the australopithecines were an offshoot that were not on the line leading to modern humans. In fact, the Leakeys argued that members of the two genera lived side by side, a proposal that has since been strengthened by considerable evidence.

During the early 1960s, after 30 years of being scoured, Olduvai Gorge began to reveal bits and pieces of a new hominid. Based on the increased size of the braincase and certain other characteristics, the Leakeys concluded that these hominids were not australopithecines and named them *Homo habilis* ("the handy man"), implying that these hominids were responsible for making the primitive stone tools found at the bottom of the gorge. Their conclusion was supported by radiodating studies that showed that these fossil remains were 1.75 million years old. The age of humans had just been pushed back in time from approximately 500,000 years for *H. erectus* to nearly 2 million years for *H. habilis*.

The argument over whether or not these specimens were actually humans (as opposed to australopithecines) continued for many years. It was finally settled to the satisfaction of most hominid specialists in 1972, with the discovery of a well-preserved skull of the same time period that was unambiguously *Homo*. The discoverer was, appropriately enough, Richard Leakey, the son of Louis and Mary, who had only recently become immersed in the hominid-hunting family passion. A body of evidence now suggests that *Homo habilis* was indeed walking the earth approximately 2 million years ago, probably in the company of several different species of australopithecines. But what type of ancestor had given rise to *Homo habilis*?

THE DISCOVERY OF LUCY

Our best clue as to the nature of that ancestor came in 1974, when Donald Johanson of the Cleveland Museum of Natural History was searching for hominid fossils at a remote site

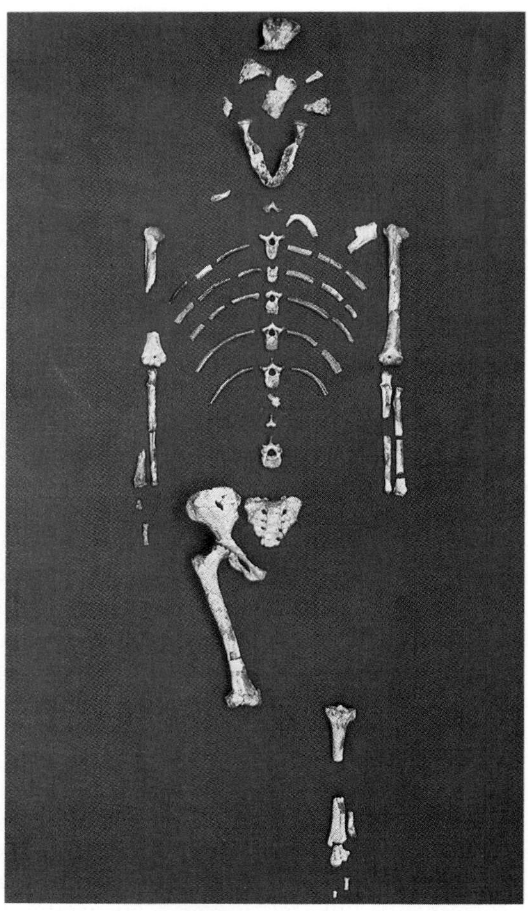

FIGURE 34-10
"Lucy," nearly 40 percent intact, has been assigned to the species *Australopithecus afarensis.* Lucy and other members of "The First Family" may represent a species of hominid that gave rise to the various other species of *Australopithecus* as well as to *Homo.*

in northern Africa, known as Hadar. Like Olduvai, Hadar was an ancient lake bed. Inhabitants of the area had died and become buried by sediment, only to be unearthed at a later time by torrents of water rushing through newly formed gullies. One morning, just as he was planning to return to camp, Johanson noticed a bone projecting out of the ground. On closer inspection, he identified it as the arm bone of a hominid. Nearby, he saw parts of a skull and a thigh bone. In his words, "An unbelievable, impermissible thought flickered through my mind. Suppose all these fitted together? Could they be parts of a single, extremely primitive skeleton?"

That is exactly what they were. Within a few weeks, all of the bones had been recovered. Together they constituted approximately 40 percent of the skeleton of an extremely primitive, small-brained female hominid who had stood only about 3.5 feet tall (Figure 34-10). Most importantly, the skeleton indicated that the hominid had walked erect, suggesting that bipedal locomotion was one of the first humanlike traits to evolve. Radiodating techniques established the age of the hominid to be 3.5 million years. She was the oldest, most complete, and best-preserved hominid fossil that has yet to be recovered. Johanson named her "Lucy," after the Beatles' song "Lucy in the Sky with Diamonds," which was playing in camp on the night of the discovery. The next year, the remains of 13 additional members of the species were found at a nearby site and have become known as "The First Family."

Johanson puzzled over the species name he should assign to Lucy and The First Family. The hominids were too primitive to be considered "human" and assigned to the genus *Homo,* particularly since there was no evidence that they had used tools. At the same time, Johanson was hesitant to place them into the genus *Australopithecus* since he considered Lucy's species an ancestor of modern humans and the australopithecines were seen by many as an evolutionary offshoot. Johanson could have established a new genus name, but that might have made matters even more controversial and confusing. He finally settled on the name *Australopithecus afarensis* (after the Afar region of Ethiopia, where the species had been discovered) and proposed that it was a common ancestor of the other australopithecines and humans. Others have argued that Lucy's species was not the only one present in east Africa 3.5 million years ago and that some of the fossils from that period are those of the genus *Homo* as well. Johanson's proposal for the evolutionary relationships among the various known species of hominids is illustrated in Figure 34-11.

What about information in the opposite direction? Can Lucy be traced back to an even more primitive ancestor? Since estimates from molecular biology suggest that the common ancestor of humans and chimpanzees lived no longer than 5 million to 7 million years ago, *A. afarensis* would not be very far removed from this common ancestor (Figure 34-11). This contention is supported by the fact that Lucy's arms are particularly long, relative to her legs, suggesting that her species had not yet lost this characteristic of an arboreal (tree-dwelling) ancestor. Somewhere in this period of a few million years, hominids had evolved the anatomic features (changes in the skull, pelvis, and leg bones) that allowed them to walk erect and to use their hands for increasingly complex activities.

These are just some of the highlights of our search for a better understanding of our origins. Only time will tell whether or not the interpretations of the few precious fossils described in these pages will hold up or will be replaced by new proposals that more accurately describe our evolutionary roots. We have just one more "fossil" to describe, not because it tells us what our ancestors looked like but because it provides a mental picture of an event that occurred one rainy day in east Africa nearly 4 million years ago. The fossil in question is a trail of footprints (Figure

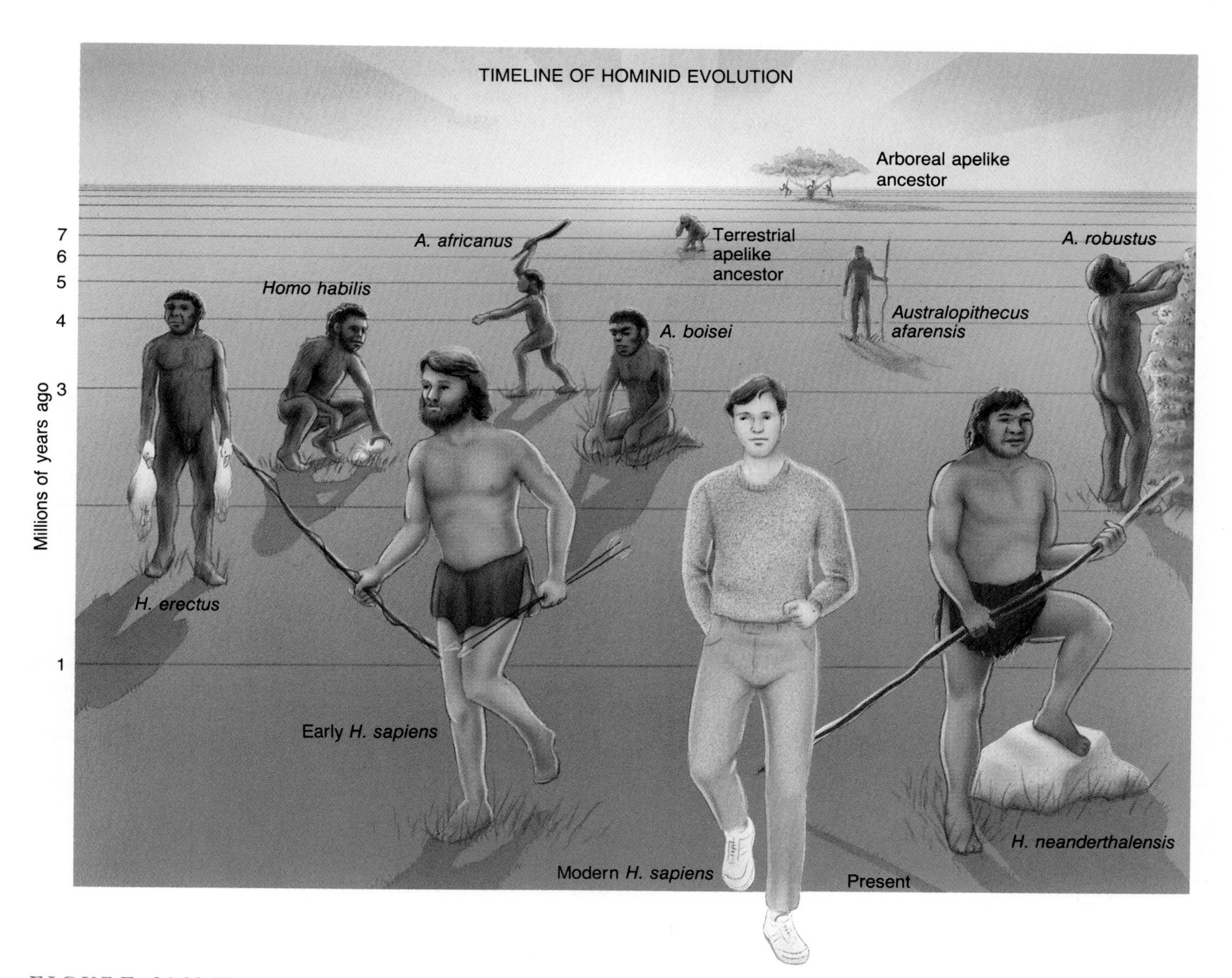

FIGURE 34-11

The human family. According to this scheme, the earliest hominid was *A. afarensis,* represented by Lucy and the First Family. Lucy walked erect and stood about 3.5 feet tall. She had an apelike skull that housed a brain about 500 cubic centimeters in volume, only slightly larger than that of similar-sized apes. She had exceptionally long arms and showed no evidence of having used tools. This species gave rise to the other australopithecines as well as the genus of humans *(Homo).* Three species of australopithecines are depicted in the illustration. *A. africanus* had the slightest build and was probably primarily a carnivore. *A. robustus* was the most heavily boned species, with huge (nickel-sized), thickly enameled molars, suggesting that they lived primarily on a diet of coarse vegetation. *A. boisei* was an australopithecine discovered by Mary Leakey, which helped focus attention on Africa as the cradle of human evolution. The first humans, *H. habilis,* appeared about 2 million years ago. *H. habilis,* like *A. afarensis,* was very small in stature and had unusually long arms, reflecting the arboreal habits of their ancestors. Their brain size was about 700 cubic centimeters, and they were able to make simple tools. *H. erectus,* which was very advanced over its predecessors, may have stood as tall as modern humans, had arms of shorter proportion than its ancestors, had a brain capacity of about 850 to 1,100 cubic centimeters, and was able to make sophisticated tools, including hand axes. There is evidence that these hominids hunted in groups and cooked their prey. While originating in Africa more than 1.6 million years ago, *H. erectus* spread over much of Europe and Asia. By about 500,000 years ago, fossils appear that are intermediate between *H. erectus* and *H. sapiens.* By 300,000 years ago, fossils are found which are unmistakably *H. sapiens.* These early *H. sapiens* had larger brains (1,200 to 1,400 cubic centimeters) and used more sophisticated tools. Modern humans *(H. sapiens sapiens)* arose between 100,000 and 200,000 years ago and are thought to have coexisted with another group of *H. sapiens,* the Neanderthals *(H. sapiens neanderthalensis),* who disappeared about 35,000 years ago.

FIGURE 34-12
Footprints made by a pair of hominids walking together across a plain of wet ash in eastern Africa approximately 4 million years ago.

34-12) that were made by a pair of fully erect hominids walking together through ash that had been spewed from a nearby volcano and then dampened by a light rain. The wet ash quickly hardened like cement and was covered by additional layers of ash, preserving the footprints (and even a few small craters left by the falling raindrops) until their discovery in 1976 by a team led by Mary Leakey. In the words of Tim White, a member of the team: "They are like modern human footprints. There is a well-shaped modern heel with a strong arch and a good ball of the foot in front of it. . . . to all intents and purposes those . . . hominids walked like you and me."

The fossil remains of hominids are fragmentary, leaving questions concerning the path that led to the evolution of modern human beings. At the same time, these remains establish without any doubt the fact that creatures—unlike any living today, with apelike faces and humanlike jaws — roamed the earth over a million years ago, walking erect and using tools. (See CTQ #4.)

REEXAMINING THE THEMES

Relationship between Form and Function

Most fossils are a structural remnant of an organism that lived many years ago. Yet, they tell us more than just how the organism was constructed; they reveal many aspects of the organism's function as well. For example, the discovery of a leg bone or skull tells us whether a primate walked on two legs or four. Similarly, the imprint of *Archaeopteryx's* wings and feathers tells us that this creature was capable of flight, but the lack of a bony breastbone indicates that the flight muscles were poorly developed, suggesting that this primitive bird probably used its wings mostly for gliding. The well-developed claws that were present on the "fingers" as well as the toes suggest that the animal may have been able to climb trees, using its clawed "fingers" to dig into the bark.

Unity within Diversity

The study of evolution encompasses the analysis of many diverse lineages that led to the appearance of many millions of different species. While the organisms generated by evolution are diverse, the underlying mechanisms are similar. Consequently, our understanding of various evolutionary pathways can be obtained by similar approaches, including the study of fossils, comparative anatomy, comparative embryology, molecular biology, and biogeography. Regardless of the particular group whose evolution is being studied, whether humans or gymnosperms, in most cases, each of these approaches can be used to provide relevant data.

Evolution and Adaptation

Evidence for the occurrence of biological evolution is revealed in virtually every area of the biological sciences. We see it when we compare the embryos of animals whose adult forms are highly diverse or when we examine structures that are vestigial and functionless in one organism but well-developed and functional in a related organism. We see it in our own bodies, where virtually every part is homologous to a corresponding part of a chimpanzee or a gorilla. We see it when we compare the types of organisms that live in a particular region of the world to similar organisms that live elsewhere. We see it in the increasing divergence of amino acid and nucleotide sequences among organisms who are thought to be more distantly related. The most direct evidence comes from studying the fossil remains of ancient organisms themselves.

SYNOPSIS

Evolution is the greatest unifying concept in biology. While biologists argue over which mechanisms may have been most important in the evolution of a particular group, there is virtual agreement that all living organisms have arisen by evolution from a common ancestor. The theory of evolution is supported by a mass of evidence gathered from several distinct biological disciplines; no credible scientific evidence has been obtained to suggest that evolution has not occurred.

When using similar characteristics to determine evolutionary relationships, care must be taken to distinguish between homologous features (those that are inherited from a common ancestor) and analogous features (those that result from convergent evolution among unrelated animals that have a similar mode of existence). Homologous and analogous features can be distinguished by applying a number of criteria, including (1) resemblance in detail, (2) similar position in relation to neighboring structures or organs, (3) similarity in embryonic development, and (4) agreement with other characters. Using these criteria, the wings of a bat and the arms of a human are homologous features, while the paddlelike limbs and streamlined bodies of fishes and whales are analogous features.

The evidence for evolution is based on the following: the study of the anatomy of living organisms, comparative embryology, the fossil record, biogeography, and biochemical and molecular data. Comparing the structure of parts of the bodies of different organisms provides evidence of evolutionary relationships. Vestigial structures (ones with little or no apparent function, such as your appendix) are considered to be remnants of structures that had a function in ancestral species that are in the process of evolutionary disappearance. Fossil analysis provides information on the types of organisms that lived in the past and evidence of the pathways by which various groups might have evolved. Comparisons of the embryos of animals can reveal homologies that are not apparent in the adults. Homologies among organisms are also revealed by comparing nucleotide and amino acid sequences of different organisms. For the most part, the longer the amount of time that has passed since two species have diverged from a common ancestor, the greater the number of substitutions found in corresponding genes and proteins between the two species. Geographic information about organisms also helps determine evolutionary relationships. Plants and animals living in nearby areas are more likely to be related than are those living far apart.

Our current view of the evolution of humans from an ancestor common to both apes and humans is based on fragmentary fossil evidence. According to this view, the first known hominid is *Australopithecus afarensis*, represented by Lucy and The First Family, who lived about 3.5 million years ago. These hominids were small in stature, had small brains, and showed no evidence of having used tools, but they walked erect and had jaws with humanlike features. *A. afarensis* may have given rise to a number of other australopithecines as well as to humans (genus *Homo*). The first known humans (*H. habilis*) appeared in Africa about 2 million years ago, and members of the species *H. erectus* appeared about 1.5 million years ago. *H. erectus* survived for over a million years, migrating to diverse regions of the earth. Modern humans (*H. sapiens sapiens*) date back 100,000 to 200,000 years.

Key Terms

hominid (p. 742)
homology (p. 744)
homologous feature (p. 744)
analogous feature (p. 744)
homoplasy (p. 744)
fossil record (p. 746)
vestigial structure (p. 749)

Review Questions

1. Name the major anatomic characteristics that underwent change during the course of human evolution over the past few million years. How are these characteristics illustrated in various fossil hominids?
2. What criteria are used to distinguish homologous and analogous features?
3. Why are vestigial structures evidence for the occurrence of evolution?
4. Compare and contrast Java Man, Peking Man, Taung Child, and Piltdown Man; arboreal and ground-dwelling hominid adaptations; primitive and derived traits; *Archaeopteryx* and fossil reptiles.
5. Why was increased brain size naturally selected during the evolution of modern humans?

Critical Thinking Questions

1. The criteria used to distinguish *Homo* from other genera, such as *Australopithecus,* have never been universally accepted. Is there one (or a few) criterion, that you feel should be the most important determinant of a "human" hominid?
2. Explain how both analogous and homologous structures provide clues to evolution. Why are homologous structures used to infer common ancestry, but analogous structures are not?
3. List one piece of evidence from each of the categories listed below that links humans with apes and/or other primates: comparative anatomy, fossils, vestigial structures, comparative embryology, comparative biochemistry.
4. Which of the four proposed family trees for hominids in the diagram below most closely resembles that described in this chapter? Why are so many different schemes proposed by different scientists working in this field?

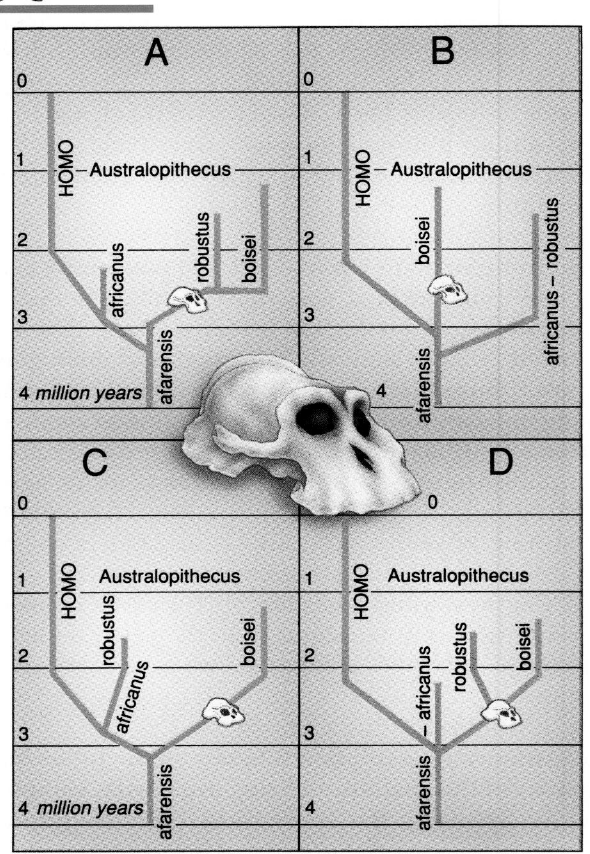

Additional Readings

Avise, J. C. 1989. Nature's family archives. *Nat. Hist.* (March: 24–27). (Intermediate)

Gould, S. J. 1980. *The panda's thumb.* New York: Norton. (Intermediate)

Gould, S. J. 1991. *Bully for brontosaurus.* New York: Norton. (Intermediate)

Lewin, R. 1987. *Bones of contention: Controversies in the search for human origins.* New York: Simon & Schuster. (Introductory)

Johanson, D., and J. Shreeve. 1989. *Lucy's child: The discovery of a human ancestor.* New York: Morrow. (Introductory)

Leakey, R. E. 1983. *One life: An autobiography.* Salem, NH: Salem House. (Introductory)

Shreeve, J. 1993. Human origins. *Discover* Jan:24–28. (Intermediate)

Wilson, A. C. 1985. The molecular basis of evolution. *Sci. Amer.* Oct: 164–173. (Advanced)

CHAPTER

35

The Origin and History of Life

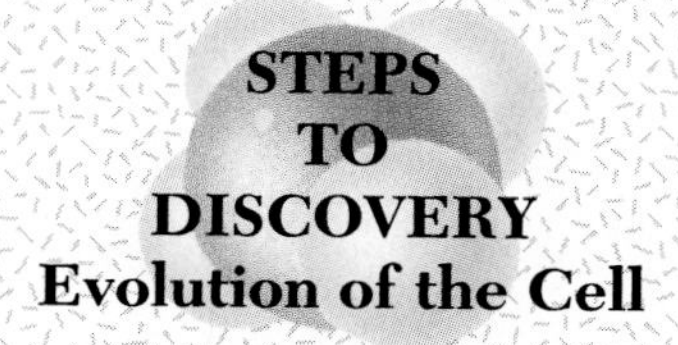

FORMATION OF THE EARTH: THE FIRST STEP

THE ORIGIN OF LIFE AND ITS FORMATIVE STAGES

THE GEOLOGIC TIME SCALE: SPANNING EARTH'S HISTORY

The Proterozoic ERA: Life Begins

The Paleozoic ERA: Life Diversifies

The Mesozoic ERA: The Age of Reptiles

The Cenozoic ERA: The Age of Mammals

BIOLINE

The Rise and Fall of the Dinosaurs

Steps to Discovery
Evolution of the Cell

When Louis Pasteur's research finally laid to rest the idea that living organisms could arise from inanimate materials (Chapter 2), he settled one nagging controversy in biology and began a new one. If life could only arise from life, how did living organisms *initially* appear on the planet?

This question was first tackled in the 1920s by the Russian biochemist Aleksandr Oparin, who proposed that life could not have arisen in a single step but only over a long and gradual process of **chemical evolution**—the spontaneous synthesis of increasingly complex organic compounds from simpler organic molecules. In Oparin's view, the formation of life occurred in several distinct stages.

The first step was the formation of simple molecules, such as ammonia (NH_3), methane (CH_4), hydrogen cyanide (HCN), carbon monoxide (CO), carbon dioxide (CO_2), hydrogen gas (H_2), nitrogen gas (N_2), and water (H_2O) during the formation of the earth's crust and atmosphere. Free molecular oxygen (O_2), which might have destroyed the earth's first delicate life forms (page 174), did not appear until the first photosynthetic organisms had evolved hundreds of millions of years later.

The second step involved the spontaneous interaction

Within the primordial seas, the first life forms appeared over 4 billion years ago by the process of chemical evolution.

of these simple molecules to form more complex organic molecules, such as amino acids, sugars, fatty acids, and nitrogenous bases—the building blocks of the macromolecules that characterize life as we know it today. The energy needed to drive the formation of these organic molecules was derived from various sources, including the sun's radiation, electric discharges in the form of lightning, and heat that emanated from beneath the earth's crust. As these organic compounds accumulated to higher concentrations in the shallow lakes and seas that dotted the primitive earth, they formed an "organic soup," in which additional reactions could take place. Some of the simpler organic molecules polymerized to form macromolecules consisting of chains of subunits, similar in basic structure to proteins and/or nucleic acids. As we discussed in Chapter 15, the first nucleic acids are thought to have been made of RNA rather than DNA.

Molecules eventually arose that could accomplish functions that were necessary for life. In this step, some of these molecules catalyzed certain chemical reactions, forming the basis for a primitive form of metabolism. Others had the capacity for self-replication and somehow formed copies of themselves. Insoluble lipids that were formed in the organic soup were forced together by hydrophobic interactions (page 59), forming lipid complexes, including membranous walls made up of lipid bilayers. Some of these self-assembling membranes formed around clusters of macromolecules, encapsulating them and forming "precells."

In the final step, precells concentrated organic molecules, allowing the molecules to react more frequently. Within the precell, catalysts directed the synthesis of specific organic polymers; the precell conducted metabolism. The eventual evolution of a genetic code enabled the precell to pass on naturally selected codes for metabolism, allowing it to reproduce itself. At this point, the precell possessed three of the basic characteristics of life—metabolism, growth, and reproduction—crossing the line between precell and living cell.

Acceptance of Oparin's theory of chemical evolution received a boost in the early 1950s as a result of a series of experiments by Stanley Miller, a graduate student at the University of Chicago. Miller demonstrated that many of the simpler compounds characteristic of life could be produced in the laboratory under conditions that were thought to have existed soon after the formation of the earth. In a sealed glass vessel, Miller repeatedly jolted a mixture of hydrogen gas, ammonia, methane, and water vapor with electric discharges in order to simulate lightning strikes. Within a matter of days, a number of organic compounds appeared in the reaction vessel, including several amino acids commonly found in proteins. Soon after, Juan Oro of the University of Houston showed that more complex biochemicals, including the nitrogenous bases that form the building blocks of nucleic acids, could be formed under similar conditions.

In 1969, support for the idea that organic compounds could be synthesized under *abiotic* (nonbiological) conditions came from an unlikely source. That September, a series of soniclike booms were heard throughout the town of Murchison, Australia, as a huge meteorite fell from outer space, exploding into pieces. Analysis of the meteorite fragments revealed the presence of a large variety of organic compounds, including amino acids, pyrimidines, and molecules resembling fatty acids. The fact that these compounds could appear under abiotic, extraterrestrial conditions made it even more likely that similar compounds had been able to form on the primitive earth.

It is one of the greatest detective stories of all time. There were no eyewitnesses, and the detectives did not come upon the scene until long after the incident had occurred. Although a host of clues had been left behind, they had all become modified over the years; virtually everything had changed.

What is this baffling case? It is nothing less than the origin and history of life on earth. The detectives in this investigation are scientists from the fields of astronomy, geology, chemistry, physics, and biology. After years of study, the *origin* of life on earth is no less a mystery today than it was decades ago. In contrast, the *history* of life on earth, from the formative beginnings of simple, prokaryotic cells to the present bewildering complexity, is becoming better understood all the time. The story begins nearly 4.6 billion years ago with the formation of the earth.

▼ ▼ ▼

FORMATION OF THE EARTH:THE FIRST STEP

Billions of years ago, our solar system was part of a massive cloud of interstellar gases and cosmic dust (Figure 35-1). As particles of matter were pulled together by gravitational attraction, the vast cloud condensed into a gigantic, spinning disk. Nearly 90 percent of the matter gravitated to the center, causing temperatures to rise high enough to ignite thermonuclear reactions. It was in this scorching center that our sun began to shine.

At the same time the sun was forming, nearly 5 billion years ago, smaller eddies of leftover gases and dust were condensing into the planets of our solar system. Heavier elements, such as iron and nickel, sank inward to form the cores of the planets, while light gases floated near the surface. Most of the lighter gases in the planets nearest the sun (Mercury, Venus, Earth, and Mars) were blasted away when the sun's thermonuclear reactions began, leaving shrunken, dense planets with virtually no gaseous atmospheres.

FIGURE 35-1
Formation of our solar system.

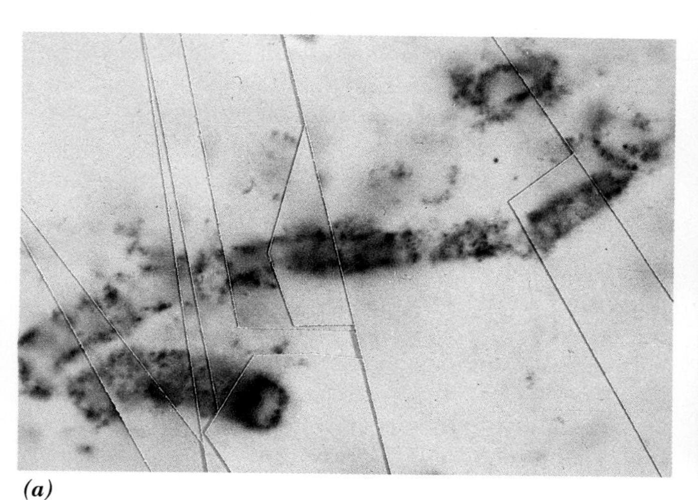

(a)

(b)

FIGURE 35-2

Imprints of the world's most ancient life. ***(a)*** A 3.5-billion-year-old cast of a filamentous cyanobacterium from western Australia. ***(b)*** These stromatolites, which dot the shore in western Australia, consist of dense masses of prokaryotic cells and mineral deposits. Prokaryotic cells have been found in stromatolites that are 3.5 billion years old.

Immense quantities of thermal energy from gravitational contraction, radioactive decay, meteorite impacts, and solar radiation turned the primitive earth into a red-hot, molten orb. As time passed, collisions with meteorites became less frequent, and the heat from radioactive decay and the earth's contraction lessened. The earth's surface slowly cooled, and a thin crust of crystalline rock formed. Below the crust, the enormous heat of the earth's interior produced massive buildups of hot gases, sparking violent volcanic eruptions that thrust molten rock and gases out through the crust. Repeated eruptions gradually built the earth's rugged land masses and filled the once empty atmosphere with clouds of hot gases and steam, first collecting into small ponds and later forming the earth's oceans, lakes, rivers, and streams. It was in these bodies of water that life emerged.

The size of the earth and its position relative to the sun were ideal for setting the stage for life, as we know it, to appear. Had the earth been much smaller, it would have lacked the mass necessary to generate enough gravitational force to hold onto its atmospheric gases. Had the earth been much closer to the sun, the scorching temperatures on the planet's surface would have prevented the condensation of steam into liquid water. Conversely, had the earth been much farther from the sun, the freezing temperatures would have kept any water in a solid, frozen state.

The first step in the formation of life on earth was the formation of the earth itself, an event that began more than 4.5 billion years ago with the condensation of cosmic matter. (See CTQ #2)

THE ORIGIN OF LIFE AND ITS FORMATIVE STAGES

In the opening pages of this chapter, we saw that the building blocks of macromolecules are readily formed abiotically, both in the laboratory and on extraterrestrial bodies. But no one has demonstrated that these simpler molecules can spontaneously assemble into reaction-catalyzing proteins or information-containing nucleic acids by the process of chemical evolution outlined by Oparin and others. Since it is impossible to recreate the origin of life in the laboratory, many different hypotheses have been offered to explain the course of events that created life. Each of these hypotheses is controversial and has been met with skepticism by some members of the "origin of life" scientific community.[1]

The first fossilized evidence of living cells is found in rocks from Australia and South Africa which date back 3.5 billion years. These rocks reveal the presence of prokaryotic cells that are not noticably different in appearance from prokaryotes living today. One such fossil consists of chains of cells (Figure 35-2*a*) that resemble modern photosynthetic cyanobacteria. Others consist of rocks formed from *stromatolites*—dense masses of bacteria and mineral deposits that grow today in warm, shallow seas (Figure 35-2*b*). The fact that such "advanced" prokaryotic cells had already appeared 3.5 billion years ago suggests that the process of chemical evolution that led to the first life forms took place relatively rapidly, probably within the earth's first 600 mil-

[1] Those interested in reading about some of these proposals might consult the article entitled "In the beginning," by J. Horgan in the February 1991 issue of *Scientific American*.

lion years. For the first 2 billion years or so, only prokaryotes populated the primitive earth. The evolutionary leap from prokaryotic cells to eukaryotic cells apparently took much longer than did the formation of the first prokaryotic cells themselves. The evolution of single-celled eukaryotes about 2 billion years ago paved the way for the evolution of the first multicellular eukaryotic organisms a billion years later. The multicellular plan was a tremendous success, leading to a proliferation of complex life forms, some of which moved onto the land and eventually took to the air.

Today, more than 2 million species of organisms populate the earth; this is only a fraction of the total number of species that have appeared over time. What caused millions of ancient species to change or to die out entirely? What were the events that constitute the history of life on earth?

Life on earth appeared in steps: Prokaryotic cells appeared first, followed by single-celled eukaryotes and, finally, multicellular eukaryotes. While fossil remnants tell us approximately when each of these stages appeared, fossils provide little information on the specific pathways by which the stages evolved. (See CTQ #3.)

THE GEOLOGIC TIME SCALE: SPANNING EARTH'S HISTORY

Geologists divide the earth's 4.6-billion-year life span according to a *geologic time scale,* which includes four great eras: the Proterozoic, Paleozoic, Mesozoic, and Cenozoic Eras (Figure 35-3). With the exception of the Proterozoic Era, the eras are subdivided into periods, which begin and end at times marked by a memorable geologic or biological event, such as a mass extinction of species. The most recent era, the Cenozoic, is further subdivided into epochs, each of which extends for a period of a few million years (see Figure 35-12).

THE PROTEROZOIC ERA: LIFE BEGINS

The **Proterozoic Era** begins with the formation of the earth's crust and stretches forward in time for approximately 4 billion years. During the beginning of the Proterozoic Era, the evolution of the first living cells took place. This stage in the history of life must have centered on sources of energy and nutrients. It is assumed that the earliest organisms were "chemical heterotrophs"—cells that were totally dependent on the amino acids, sugars, and other organic compounds that had accumulated in the "organic soup" as the result of abiotic chemical reactions. An abundant supply of nutrients probably allowed these early heterotrophs to proliferate at a rapid rate. Eventually, however, these organisms must have consumed the organic molecules faster than abiotic synthesis could replace them, causing competition for limited nutrients.

At some point, cells must have evolved simple energy-releasing pathways, such as fermentation (Chapter 9), that enabled them to extract energy from available nutrients and store it in the form of ATP. Such cells may have been similar to modern anaerobic bacteria that live in the sediments beneath ponds and oceans. But even the early prokaryotic heterotrophs that were capable of fermentation depended on organic compounds formed abiotically for their source of nutrients. The only organisms that could be freed from the limitations of abiotically synthesized organic molecules were those that could manufacture their own organic nutrients from inorganic precursors, using energy obtained from their abiotic environment. These were the earth's first autotrophs.

The Evolution of Autotrophs

If it were not for the emergence of autotrophs, life might have ended 3.5 billion years ago. The first autotrophs not only supplied themselves with nutrients, they also became food for the heterotrophs. These early autotrophs were probably very similar to present-day anaerobic, photosynthetic bacteria that live in well-lit, oxygen-deficient environments, such as stagnant ponds, marshes, and swamps. The first autotrophs used the energy of sunlight to remove electrons from hydrogen gas, hydrogen sulfide, ethanol, or lactic acid; they combined the liberated electrons (and protons) with carbon dioxide to produce organic compounds. These compounds could then be used in glycolysis and fermentation to produce ATP. The earth's deposits of sulfur (formed by the oxidation of hydrogen sulfide) are thought to be a legacy of these early bacteria.

The metabolic success of anaerobic photosynthesizers soon gave way to another group of organisms that used light energy to split water, extracting electrons and releasing molecular oxygen as a waste product. These new autotrophs were the first **cyanobacteria,** a 3-billion-year-old group of photosynthetic bacteria that continue to flourish today. Because water was an enormously plentiful resource, these ancient cyanobacteria flourished, their light-capturing photosynthetic pigments tinting the ponds and seas green. These autotrophs released vast amounts of molecular oxygen into the water and air, which posed an enormous threat to the anaerobic organisms and led to the evolution of mechanisms that protected cells from this potentially deadly substance (page 56). Organisms that were unable to cope with an aerobic environment were either driven to extinction or were forced into very limited, oxygen-free habitats.

From Prokaryotic to Eukaryotic Cells

Early eukaryotes left no fossilized remains. Furthermore, there are no organisms living today whose complexity lies between the prokaryotic and eukaryotic states. In the absence of living intermediates, speculations on how eukaryotic cells evolved from simpler prokaryotic ancestors are

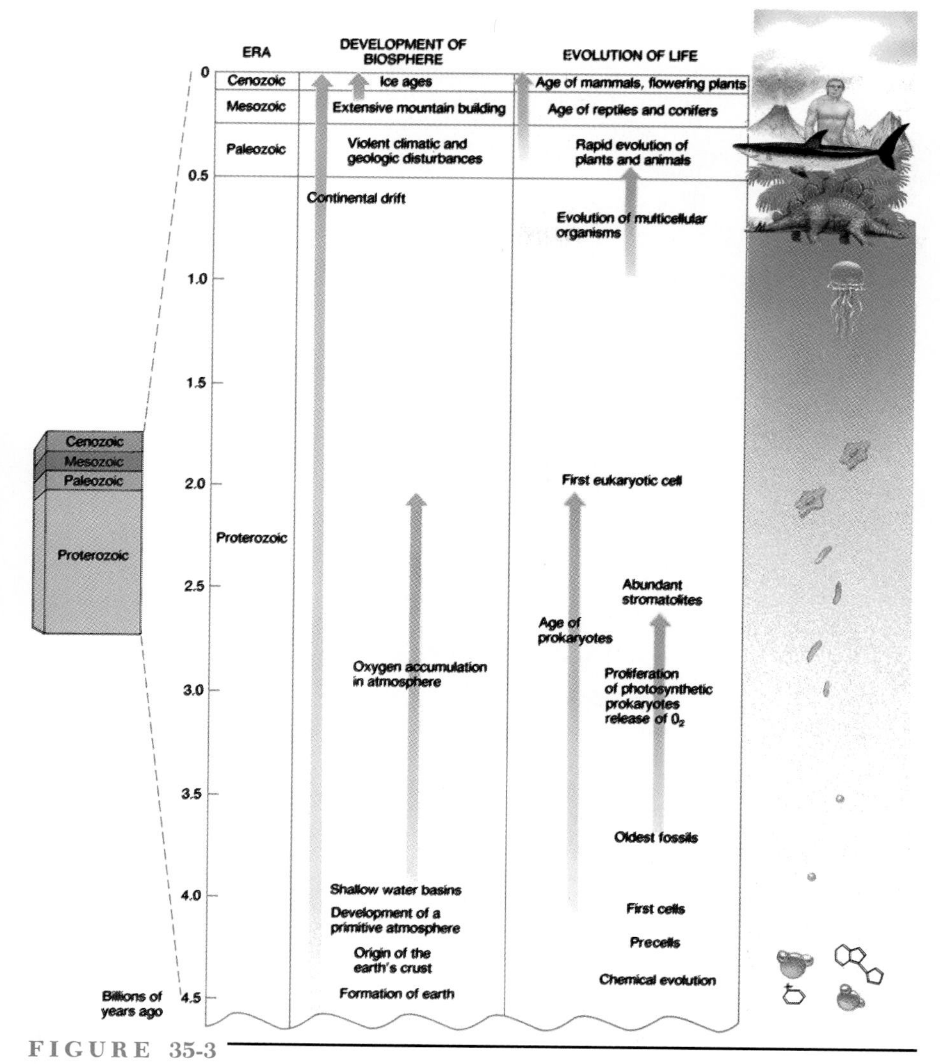

FIGURE 35-3

The geologic time scale. The 4.6-billion-year life span of the earth is divided into four eras, based on changes in the fossil record and the earth's terrain. Major changes in the physical makeup of the earth and its atmosphere dramatically affected the evolution of life.

based on the structures and functions of modern single-celled organisms. Two theories have been proposed to explain the evolution of eukaryotic cells, each of which may explain part of the story (see page 110). Some of the organelles of a eukaryotic cell, such as the mitochondria, chloroplasts, and cilia, may have evolved from small prokaryotic cells that took up residence within larger eukaryotic hosts. Other eukaryotic organelles, such as the endoplasmic reticulum and the nuclear envelope, may have arisen from membranes that invaginated from the cell surface.

From Unicellular Eukaryotes to Multicellular Eukaryotes

Multicellular life began nearly 1 billion years ago. Multicellularity is thought to have evolved independently in a number of different types of eukaryotes. In some cases, multicellularity may have been achieved when a group of independent, single-celled eukaryotes aggregated to form a colony of cells. In other cases, colonies arose when the daughter cells that had formed as a result of cell division remained together as a group of cells rather than going their separate ways. Over time, the cells in the colonies lost their capability for independent life.

Multicellularity provides the opportunity for a division of labor among the constituents involved. Gradually, some of the cells became specialized for specific functions, which they carried out with greater efficiency than did other cells in the colony. For example, some cells became specialized as locomotor structures, moving the colony from place to place; others became specialized for obtaining and processing nutrients; still others became specialized for reproduction by acquiring the capability for meiosis and gamete formation. Eventually, what had originated as a colony of eukaryotic cells had become a true multicellular plant or animal. The organization of a number of living organisms lies somewhere between the single-celled and multicellular

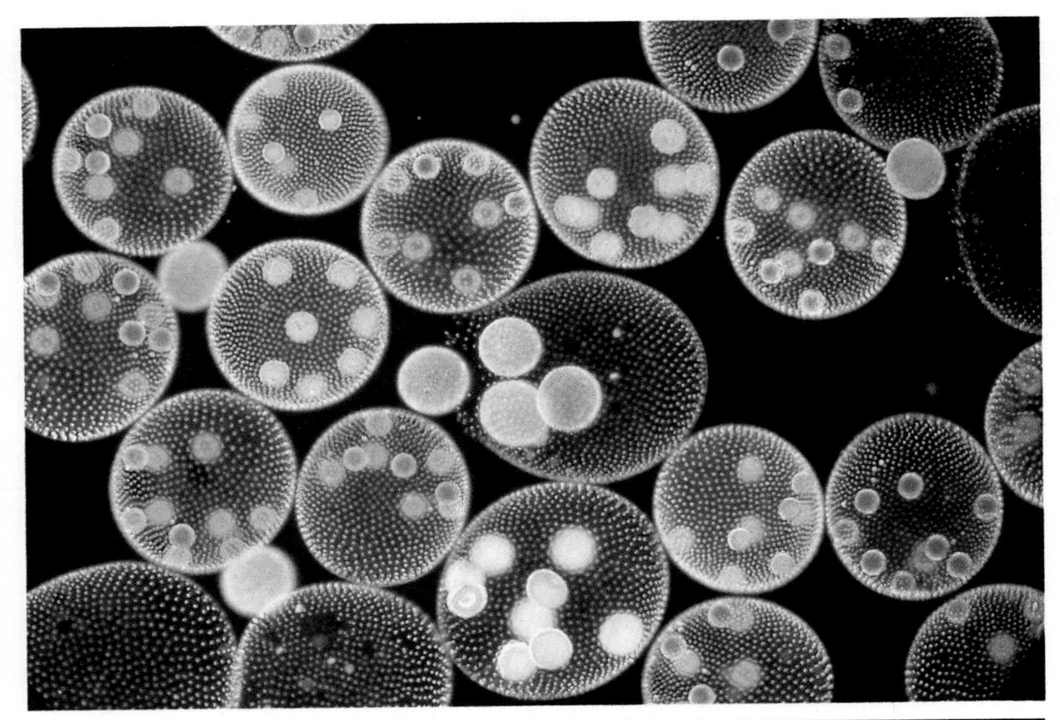

FIGURE 35-4
Volvox, a colonial protist. The cells of this colonial protist arise by mitosis from a single cell. Within the colony are cells specialized for locomotion, photoreception, and sexual reproduction. It has been proposed that a colonial organism such as *Volvox* gave rise to all multicellular organisms. The dense green masses inside the large spheres are daughter colonies that will be released into the surrounding environment.

state. The best studied is the colonial protist *Volvox* (Figure 35-4).

▶ By the time the Proterozoic Era ended, members of all the major animal phyla had appeared. We know this to be the case since at the beginning of the next era, fossils representing the various phyla are "suddenly" found in considerable numbers. The lack of Proterozoic fossils probably reflects the fact that the animals living in this era lacked hardened skeletal parts that are usually required for fossilization. A few valuable samples of fossilized Proterozoic life remain, however. The best samples come from a region of sandstone in South Australia, where a group of marine organisms, including jellyfish (cnidarians), segmented worms (annelids), and the soft-bodied ancestors of arthropods became deposited in the fine mud and silt of the Proterozoic ocean bottom about 650 million years ago. Even though the animals lacked skeletons, the imprints of their bodies were preserved in the forming sandstone (Figure 35-5).

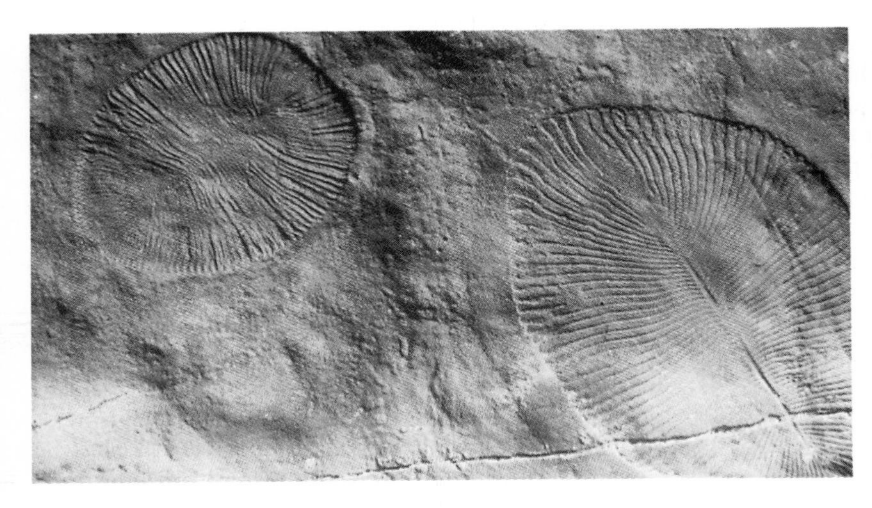

FIGURE 35-5
The imprint of a soft-bodied, multicellular animal that lived approximately 650 million years ago. The animal appears to have been segmented but lacked a head or appendages.

THE PALEOZOIC ERA: LIFE DIVERSIFIES

The **Paleozoic Era** lasted approximately 345 million years. Conditions on earth changed dramatically a number of times during this long era, causing episodes of mass extinctions, followed by diversification of organisms via adaptive radiation, whereby a single ancestor gives rise to a diverse array of organisms adapted to different habitats (page 734). Many of these changes were the results of shifts in the positions of the earth's continents. At times, the drifting continents created broad, shallow seas; at other times, land masses became entirely submerged and later rose to form mountain ranges. These changes had enormous impact on the earth's fauna and flora.

Continental Drift

In 1968, an expedition to Antarctica unearthed a collection of fossil reptiles that were virtually identical to fossils found in Africa and India, locales situated thousands of miles apart. This finding raises important questions: How can we explain the presence of the same species of terrestrial vertebrates living on several different continents? And how could these "cold-blooded" reptiles have lived in Antarctica, which is so cold that it is home to only a few species of heavily insulated, "warm-blooded" birds and mammals?

In 1912, Alfred Wegener, a German geologist, hypothesized that the continents were not always situated in the same place on the earth's surface as they are today. Wegener observed that the continents of Africa and South America, which are situated on opposite sides of the Atlantic Ocean, have complementary outlines and might be able to fit together spatially. He studied the geologic formations on the opposing coastlines and realized that many of the features that ended abruptly on one continent continued on the opposite continent. Wegener spent the remaining 20 years of his life amassing evidence for his theory that the continents were once situated close together and had then drifted apart, a phenomenon known as **continental drift.**

The occurrence of continental drift is now explained by the theory of *plate tectonics.* According to this theory, the solid crust of the earth, which is estimated to be about 100 kilometers (62 miles) thick, consists of a number of rigid plates (Figure 35-6) that rest on an underlying layer of

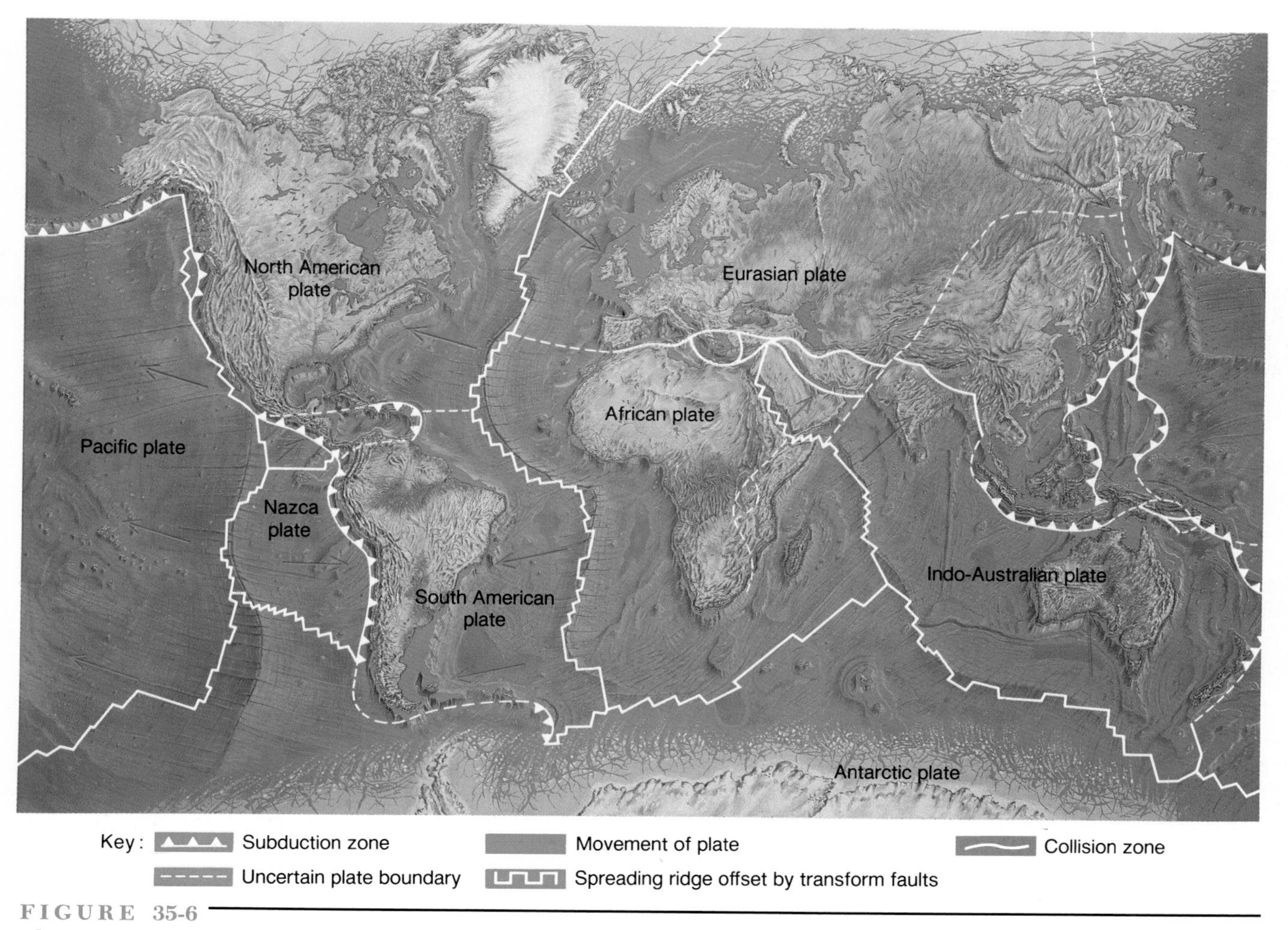

FIGURE 35-6
The geographic distribution of the major tectonic plates that make up the earth's crust.

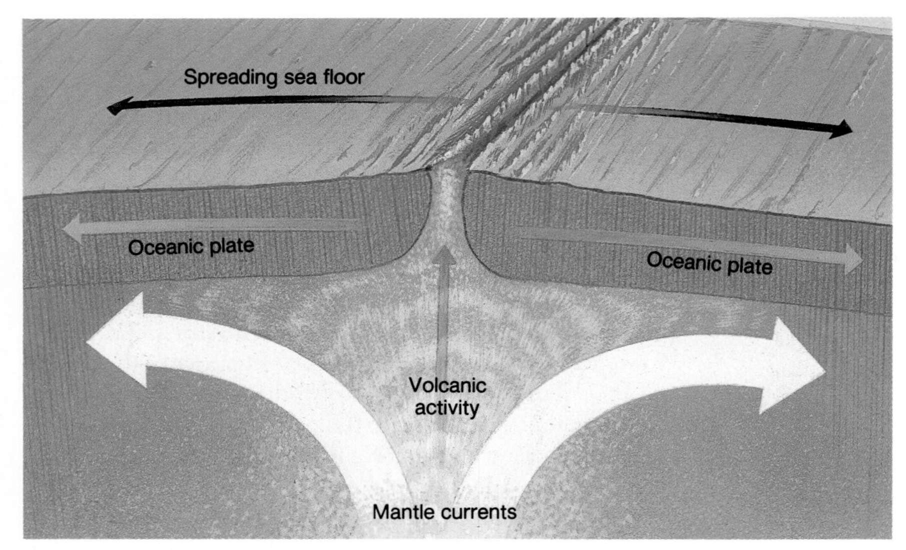

(a)

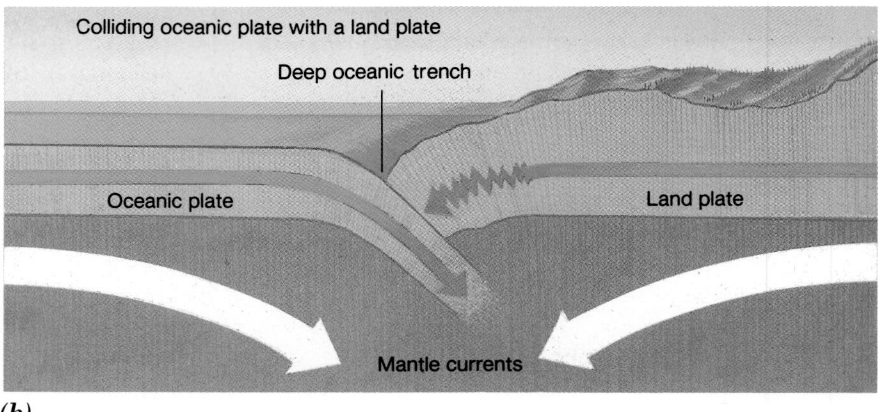

(b)

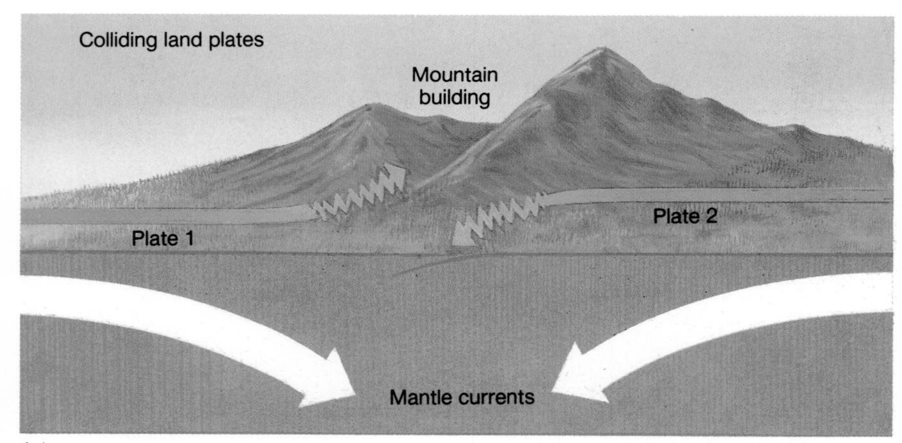

(c)

FIGURE 35-7

The basis of continental drift. ***(a)*** The plates that make up the earth's crust are in a state of dynamic interaction. When magma rises up from the earth's core, the molten material becomes hardened as part of one of the earth's plates. This process pushes the plate outward toward a collision with another plate at its edge. ***(b)*** In the case depicted here, the edge of an oceanic plate is sliding under a land plate (one bearing a continental land mass). The Pacific and North American Plates, for instance, are sliding toward each other, causing shifts in land and repeated earthquakes along the San Andreas Fault, the line where the two plates "sideswipe" each other. ***(c)*** If two plates crumble when they collide, mountain ranges are built. The Andes, Alps, and Himalayas were formed in this way.

semimolten (hot, semiliquid) rock. The movement of the earth's plates is very gradual, from 1 to 10 centimeters per year, and results from the movement of molten rock (*magma*) upward into the solidified crust along certain ridges within the ocean floor (Figure 35-7). The incorporation of new material at the ridges expands the ocean floor, pushing the edges of the plates outward. The continents move passively, as the underlying plates are pushed along. As adjacent plates are forced together, the edge of one plate slides downward into the oceanic trench (is *subducted*), where it is taken back into the earth's core and reconverted to molten rock. The pressures generated by the collision of crustal plates is a cause of earthquakes and the uplifting of mountains. For example, the 1989 earthquake that caused major destruction in the San Francisco bay area, was a result of the Pacific and North American plates sliding toward each other. Similarly, the Himalayas were formed in the wake of the collision between Asia and the Indian subcontinent 50 million years ago. Events of such magnitude have enormous impact on organisms living in the affected habitats.

According to present data, the continental portions of the earth's crust were clustered together 200 million years ago, forming a "supercontinent" known as Pangaea (Figure 35-8). Approximately 150 million years ago, Pangaea began to fragment, initially into two large continents, Laurasia and Gondwanaland, causing the separation of many populations of organisms. Over time, these two large continents split apart; the resulting continents drifted to their present positions, carrying with them fossils of organisms that had once flourished half a world away. This is why the same species of reptiles are spread across land masses ranging from India

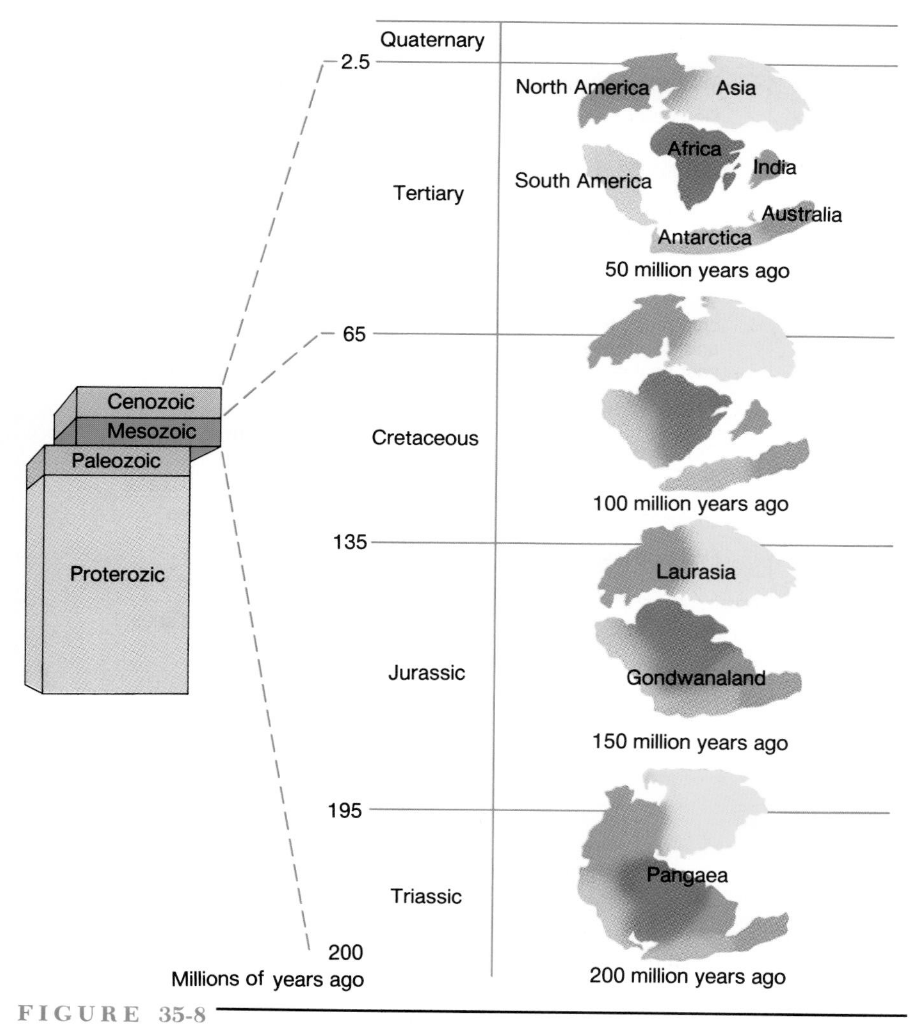

FIGURE 35-8
The consequences of continental drift. Nearly 50 million years after Pangaea formed, it began to fragment into two large continents—Laurasia and Gondwanaland—separating populations of organisms. Over the next 150 million years, these large continents split apart, and the smaller continents drifted to their present positions.

and Africa to the Antarctic; these continents were once joined together in a common land mass.

Continental drift alone does not explain how reptiles (along with a variety of plant life) were once able to thrive on the Antarctic continent, which today is covered by ice and the scrubby vegetation of the tundra. If you examine the series of maps in Figure 35-8 you will notice that Antarctica has not moved very far from its original position as part of Pangaea, yet the country has changed from a land of forests to a frozen wilderness. This transformation reflects the dramatic changes in climate that have occurred on earth over the past couple of hundred million years.

Mass Extinctions

The Paleozoic Era is divided into six periods: the Cambrian, Ordovician, Silurian, Devonian, Carboniferous, and Permian, each of which is characterized by the appearance of major groups of plants and animals (Figure 35-9). For example, jawless fishes, the first vertebrates, appeared about 475 million years ago. Over the following 200 million years or so, terrestrial habitats were colonized by a succession of multicellular eukaryotes, both plants and animals. The first primitive land plants were followed by mosses and ferns, then seed plants, and eventually gymnosperms. Among the first land animals were scorpionlike arthropods, then wingless insects, amphibians, and, eventually reptiles. The final period of the Paleozoic Era (the Permian Period) was a time of great change, both geologically and biologically. The climate appears to have become much colder and dryer, apparently culminating in a period of glaciation.

About 245 million years ago, as the Paleozoic era neared an end, life on earth experienced the most disastrous mass extinction in its entire history. This was particularly true for sea life; upward of 80 to 90 percent of all marine

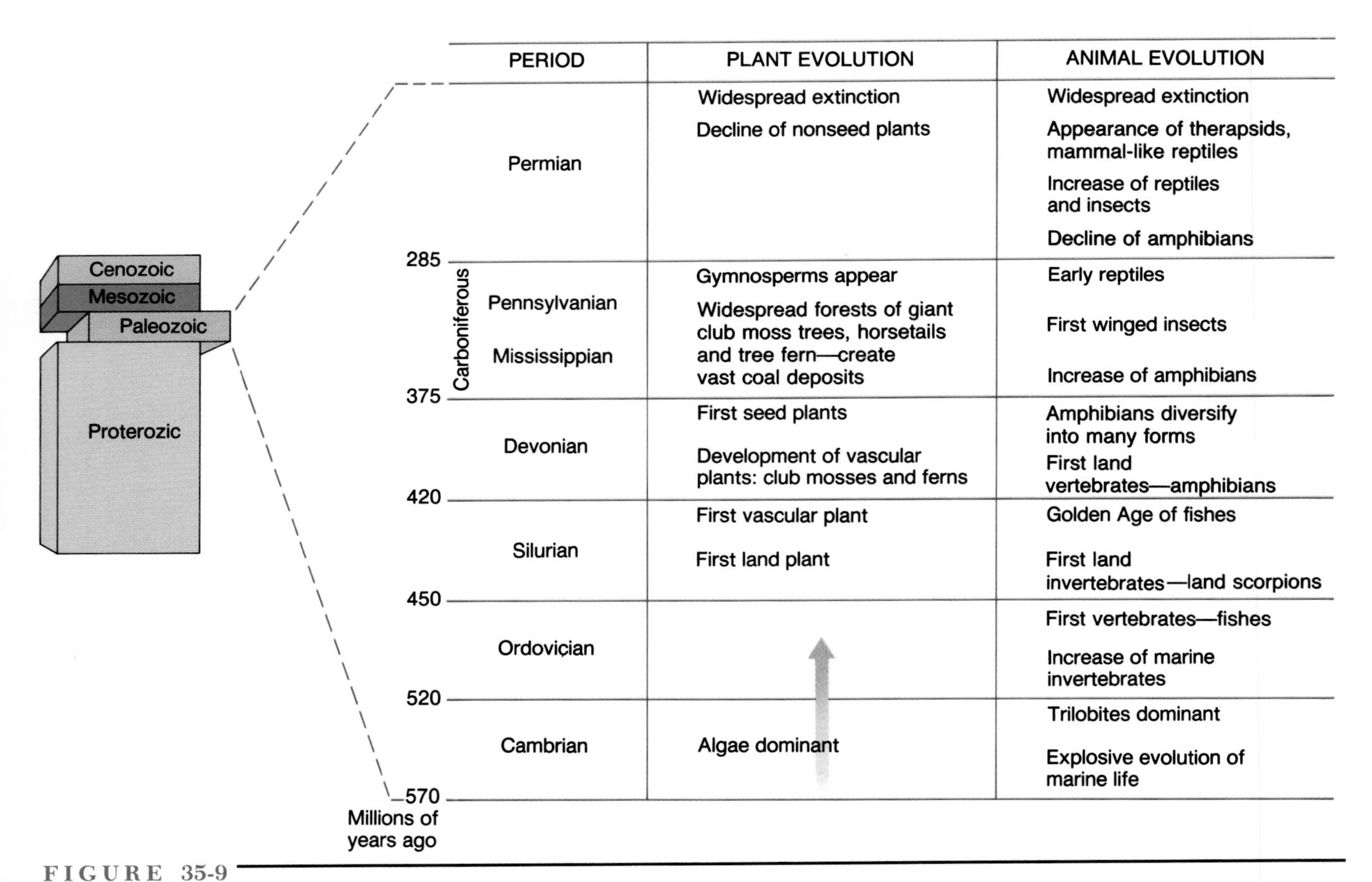

FIGURE 35-9

The Paleozoic Era. At the beginning of the Paleozoic Era, the seas were brimming with algae and primitive invertebrates, such as sponges, jellyfish, worms, starfish, and trilobites (an extinct group of arthropods). Later on, plants, and then animals, began to colonize the land. Near the end of the Paleozoic Era, forests of giant tree ferns, club mosses, seed ferns, horsetails, and early conifers (evergreen trees and shrubs) blanketed huge expanses of the earth's surface. Amphibians were the dominant land animal at this time. The end of the Paleozoic Era was marked by the greatest mass extinction ever recorded in the fossil record. Among the survivors were two groups of organisms: the conifers and the reptiles.

species disappeared from the earth. The reason for this dramatic decrease in marine organisms is unclear. There is no evidence that it was the result of some sudden catastrophe since careful examination of the fossil record indicates that numbers of species dwindled over a period of several million years. Meanwhile, on land, many amphibian and mammal-like reptile groups also disappeared, creating new opportunities for the rise of other reptiles as the dominant terrestrial vertebrates during the Mesozoic Era. Of all life forms, insects and plants appear to have been least affected by the Permian extinctions. As we will see, such episodes of mass extinction occurred during the next 2 eras as well.

THE MESOZOIC ERA: THE AGE OF REPTILES

The **Mesozoic Era** (65 million to 240 million years ago) was generally a time of stable weather patterns, extensive mountain building that produced much of the earth's present terrain, and rising seas. The Mesozoic Era (Figure 35-10) is divided into three periods, the Triassic, Jurassic, and Cretaceous.

The Mesozoic Era is sometimes called the Age of Reptiles because this group of vertebrates underwent an adaptive radiation in which diverse forms spread into all of the various terrestrial habitats, and even back into the water. One group, the *pterosaurs,* became the first vertebrates to evolve the ability to fly. Another group of reptiles evolved into the largest land animals ever to roam the earth—the dinosaurs; they would "rule" the land for 125 million years. But the reign of the dinosaurs collapsed in "sudden" extinction at the end of the Mesozoic Era. How could the dinosaurs have fallen from dominance to extinction in less than a million years? The answer still remains a mystery (see Bioline: The Rise and Fall of the Dinosaurs), but whatever caused the extinction of these powerful animals also caused the extinction of many other forms of life, including many marine organisms.

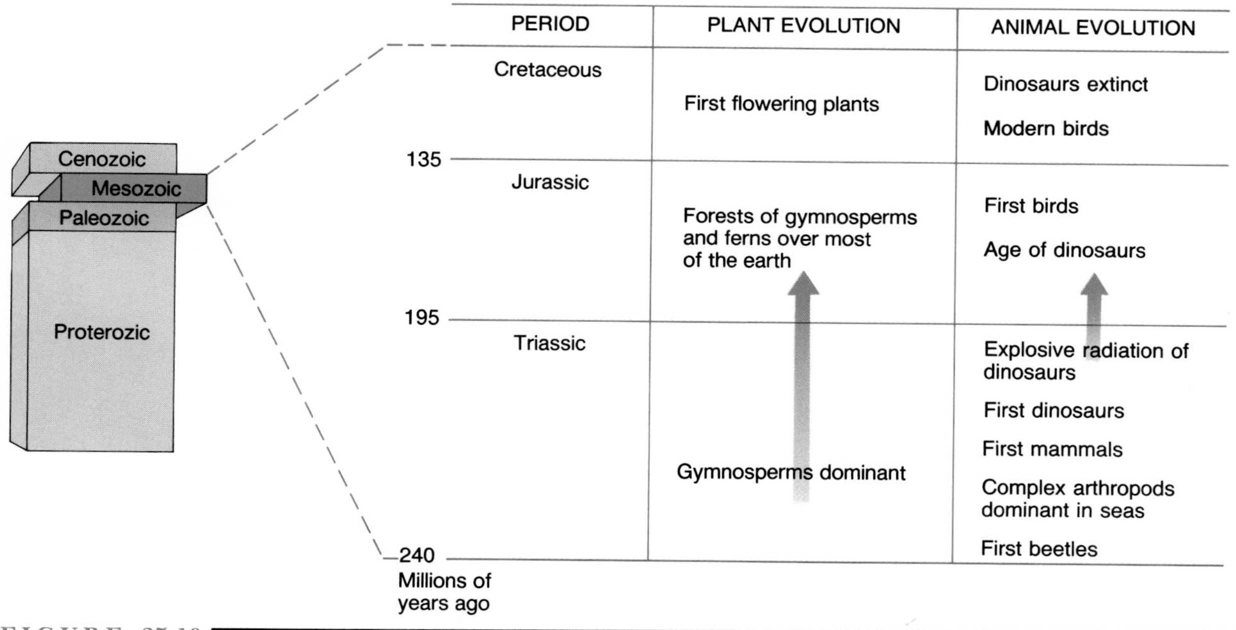

FIGURE 35-10
The Mesozoic Era. The Mesozoic Era is often called the Age of Reptiles. Reptiles underwent an adaptive radiation that filled the earth's habitats with representatives of this great class of vertebrates. Included among the Mesozoic reptiles were the dinosaurs and the therapsids, the ancestors of mammals. The first flowering plants evolved near the end of the Mesozoic Era, which, like the end of the Paleozoic Era, was marked by mass extinctions.

◁ BIOLINE ▷

The Rise and Fall of the Dinosaurs

The dinosaurs, or "terrible lizards," began their adaptive radiation about 225 million years ago, approximately 100 million years after the first reptiles had evolved from amphibians. The earliest species were relatively small carnivores with a bipedal (two-legged) gait that likely carried the animals in swift pursuit of prey. From these modest beginnings, a variety of carnivores of huge dimensions evolved. Perhaps the largest of these was *Tyrannosaurus rex* (Figure 1), which weighed over 6 tons and stood over 4.5 meters (15 feet) tall. The dinosaur's powerful jaws contained large, swordlike teeth. The animal had huge hindlimbs, but its forelimbs were so poorly developed, they were unable to reach its own mouth.

Tyrannosaurus was dwarfed compared to some of the herbivorous dinosaurs, such as *Brachiosaurus,* which weighed 80 tons and reached 15 meters in height, and *Apatosaurus* (formerly *Brontosaurus*), which reached lengths of 25 meters. The herbivorous dinosaurs had massive bodies with disproportionately small heads and brains and very long necks and tails (Figure 2). The animal's front teeth were similar in shape to our own incisors and were used to crop the plants on which the dinosaur browsed. Since they had no rear teeth, herbivorous dinosaurs were probably unable to chew their food and relied instead on some type of grinding mechanism located within their digestive tract (as in modern birds, which evolved from one of the dinosaur lines). Unlike the carnivores, these larger herbivores walked on four legs.

Reptiles dominated the land, air, and water for most of the Mesozoic Era, a reign exceeding 125 million years. Then, about 65 million years ago, the dinosaurs disappeared from the fossil record, along with a wide variety of other unrelated organisms, including many species of plants, marine invertebrates, and single-celled protists. Many hypotheses have been presented over the decades to explain the extinction of the dinosaurs and other life forms. Taken together, these hypotheses fall into two distinct categories: those that suggest a gradual process of extinction that took place over a period from 2,000 to 3 million years, versus those that propose a catastrophic extinction that might have occurred over a period of months to years. You might think that the fossil record would enable paleontologists to distinguish easily between a "gradual" versus a "sudden" course of extinction, but this has not been the case. Depending on which organisms have been scrutinized, paleontologists have argued for years over whether or not the loss of certain fossil species is consistent with one or the other hypothesis.

In 1980, a paper was published by Luis Alvarez, a Nobel laureate in physics, and his colleagues. This paper, and the research it stimulated, has gradually shifted opinion over to the side of catastrophic extinction. Alvarez and colleagues were examining the composition of the rocks that

FIGURE 1
A skull of *T. rex* emerging from South Dakota sandstone.

were formed about the time the dinosaurs became extinct, a time that separates the Mesozoic Era from the Cenozoic Era, 65 million years ago. The scientists found a thin layer of rock that contained highly elevated levels of a platinumlike element called *iridium.* Iridium is normally present at low levels in the earth's own crust and at high levels in extraterrestrial bodies, such as asteroids and meteorites. Alvarez interpreted this distinction as evidence that a huge asteroid or meteorite must have struck the earth 65 million years ago, sending a massive cloud of iridium-containing dust into the atmosphere. The dust was dispersed around the world before settling back onto the earth, forming a thin layer that has been preserved in the earth's rocks. In the past few years, corroborating evidence for such an impact has come to light from the discovery of tiny, glasslike particles called *tektites* at the Mesozoic/Cenozoic boundary. Tektites are generated by the tremendous pressure of a meteorite as it hits the ground at speeds of 15 kilometers per second. Current evidence points to the Caribbean as the most likely site of impact.

How can the impact of a meteorite in the Caribbean affect the survival of a dinosaur living thousands of kilometers away on the other side of the earth? Scientists generally agree that such an impact could have radically altered the climate on earth for a period of several years; they disagree on the type of climatic changes that resulted, however. Alvarez initially proposed that the impact would have generated a cloud of dust in the atmosphere, similar to that caused by the eruption of a large volcano. The dust would have blocked the sun's vital rays, greatly lowering the levels of photosynthesis on which virtually all organisms on earth ultimately depend.

Other possible effects of a meteorite have been suggested, including a period of elevated warming due to the greenhouse effect (resulting from increased levels of carbon dioxide in the atmosphere due to global wildfires), the formation of acid rain (due to increased levels of nitrous oxide in the atmosphere), and the release of natural gas (methane) trapped in the earth's crust. We may never know precisely what it was that killed the dinosaurs, but it remains a fascinating topic that captures the imagination of children and geochemists alike.

FIGURE 2

A skeleton of the herbivorous dinosaur, *Diplodocus*. While the accompanying human skeleton is useful in revealing the size of these dinosaurs, closer examination of the limbs, pelvic girdle, and ribcage exposes numerous homologies between the two vertebrates.

One group that was not as severely affected by extinction consisted of small, probably nocturnal animals that had evolved earlier from a group of reptiles known as *therapsids* (Figure 35-11) but had remained in the "shadows" as the reptiles underwent their diversification. These animals were the mammals, and they are thought to have had a number of important biological innovations: They maintained a constant, elevated body temperature that allowed them to remain active at night; they were covered with hair that insulated them from colder temperatures; and they gave birth to their young, rather than hatching them from eggs.

Birds, insects, and mammals survived whatever it was that caused the mass extinction of the ruling reptiles, and flowering plants emerged and thrived. Once the dinosaurs and other large reptiles had become extinct, birds and mammals quickly filled their vacated habitats, giving rise to a number of groups that would dominate animal life during the following era, the Cenozoic.

THE CENOZOIC ERA: THE AGE OF MAMMALS

The present era, the **Cenozoic Era**, began about 65 million years ago and is known as the Age of Mammals (Figure 35-12). Insulating hair and thermoregulatory abilities adapted mammals for the colder climates that became more prevalent during the Cenozoic Era. By the first 10 million to 15 million years of this era, representatives of most of the modern orders of mammals had appeared. During the first half of the Cenozoic Era, the order Primates was represented by the lemurs and tarsiers, small arboreal forms that resembled some of the earliest mammals. Monkeys appear in the fossil record approximately 35 million years ago, and the direct ancestors of the four groups of modern apes (gibbons, orangutans, chimpanzees, and gorillas) appeared about 20 million years ago. The Cenozoic Era also saw the explosive radiation of flowering plants.

The Pleistocene Epoch, which began about 2.5 million years ago, was characterized by intermittent ice ages that caused mass extinctions and migrations, remolding the distribution of life on earth. Once vastly distributed, tropical vegetation became restricted to habitats with mild climates, near the equator. During the last ice age, giant mammoths, ground sloths and numerous other large mammals became extinct (Figure 35-13). Following this last ice age, which ended about 10,000 years ago, semiarid and arid areas developed, and the surviving plants and animals began the most recent period of adaptive radiation.

Now that we have described the major trends of evolution that have occurred on earth over the past several billion years, we are ready to take a closer look at the diverse forms of life that have been generated by evolution. In the following section of the text, we will explore the major characteristics of the organisms that comprise each of life's five kingdoms.

The history of life on earth has been marked by progression, decimation, and renewal. During the first 3.5 billion years or so, the earth was the site of the evolution of increasingly complex life forms, culminating in multicellular plants and animals. The past 700 million years have seen waves of mass extinctions, which, in some cases, nearly decimated plant and animal life on earth, followed by periods in which new groups rose to prominence, replacing those that either receded in diversity or disappeared altogether. (See CTQ #4.)

FIGURE 35-11
A therapsid reptile of the type that gave rise to mammals. These animals' teeth, jaws, and "upright," four-legged posture made them more like mammals than other reptiles.

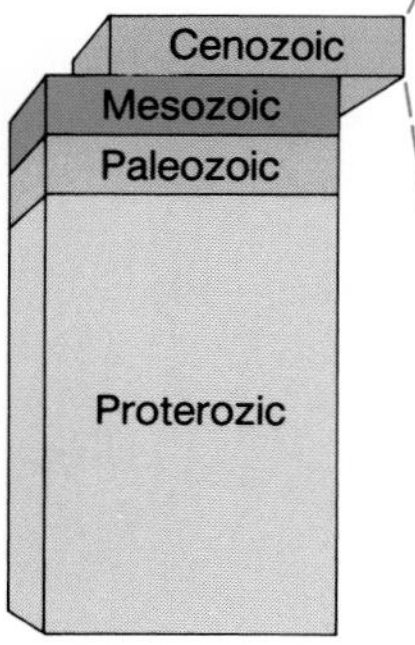

Cenozoic
Mesozoic
Paleozoic
Proterozic

PERIOD	EPOCH	PLANT EVOLUTION	ANIMAL EVOLUTION
Quaternary	Recent	Increase of herbaceous plants	Appearance of *Homo sapiens*
	Pleistocene	Repeated glaciation leads to mass extinction	Repeated glaciation leads to mass extinction First *Homo*
2.5			
Tertiary	Pliocene	Decline of forests, spread of grasslands	Appearance of hominids
	Miocene		Appearance of first apes
	Oligocene		All modern genera of mammals present
	Eocene		In seas, bony fish abound
	Paleocene	Explosive radiation of flowering plants	Rise of mammals First placental mammals
65			

Millions of years ago

FIGURE 35-12
The Cenozoic Era is divided into two periods, the tertiary and quaternary, which together are subdivided into seven epochs. Mammals and flowering plants evolved rapidly during the Cenozoic Era. Intermittent ice ages continually remolded the distribution of life and caused the extinction of many organisms. Since mammals became the dominant land animal at this time, the Cenozoic Era is also called the Age of Mammals.

REEXAMINING THE THEMES

Acquiring and Using Energy

During the earliest period of chemical evolution on earth, the energy required for the formation of complex organic chemicals from simpler compounds was derived from a number of sources, including the ultraviolet radiation from the sun, the disintegration of radioactive atoms, the impact of meteorites, molten lava from the earth's core, and electric discharges from lightning. The first cells on earth were presumably heterotrophs. They fueled their activities with the chemical energy present in the compounds that made up the organic soup in which they lived. Ultimately, autotrophs evolved that could capture energy from sunlight and convert it to the chemical energy needed to synthesize their own organic compounds.

Evolution and Adaptation

Evolution of life on earth has progressed hand in hand with evolutionary changes in the earth itself. During the first billion or so years after the earth's formation, the earth's atmosphere was very different from how it is today. It probably contained nitrogen, methane, ammonia, and hydrogen gas but was virtually devoid of molecular oxygen. This atmosphere was conducive to chemical evolution. It allowed for the abiotic formation of organic compounds and eventually led to the formation of life. By about 3 billion years ago, the cyanobacteria began filling the atmosphere with molecular oxygen, promoting the preponderance of aerobic life on earth. The evolution of life has also been markedly affected by geographic and climatic changes. Continents have come together and drifted apart, while global temperatures have varied from that of tropical environments to ice ages. Like the appearance of oxygen, these changes have had dramatic effects on the earth's flora and fauna, resulting in waves of extinction, followed by periods of great biological resurgence, in which newly evolved organisms adapted to the new prevailing conditions.

FIGURE 35-13

A gallery of Ice Age animals. From left to right: Jefferson's mammoth, Conkling's pronghorn antelope, giant heron vulture, and a great short-faced bear.

SYNOPSIS

The earth condensed out of a massive cloud of dust and gases, approximately 4.6 billion years ago. Life subsequently appeared in less than a billion years, the result of spontaneous chemical evolution. The first life forms are presumed to have been chemical heterotrophs that obtained their energy and materials from organic compounds in the surrounding medium. The evolution of prokaryotic autotrophs about 3 billion years ago was essential for the continuation of life on earth. The most advanced autotrophs were the cyanobacteria, which were able to utilize the energy in sunlight to split water, gaining electrons as reducing power and releasing molecular oxygen. Unicellular eukaryotes evolved about 2 billion years ago, paving the way for the evolution of multicellular organisms—presumably from colonial protists—about 700 million years ago.

The geologic time scale divides the earth's lifespan into four great eras, beginning with the Proterozoic Era, during which life evolved. Repeated changes in the shape, size, and location of the earth's continents drastically changed the earth's climate and sea level, affecting the evolution of organisms. By the end of the Proterozoic Era, members of all of the major animal phyla had appeared, even though the fossil record of these groups is very limited.

Vertebrates evolved during the Paleozoic Era, first in the sea and then on land. Arthropods also underwent an adaptive radiation on land, beginning with scorpionlike forms, followed by a variety of wingless and winged insects. Mosses, ferns, and seed plants appeared successively on land. The Paleozoic Era ended with the greatest episode of mass extinction of organisms to have ever taken place, extinguishing the vast majority of marine organisms and many terrestrial forms.

During the Mesozoic Era, reptiles and gymnosperms dominated the land. Adaptive radiation among reptiles produced a diversity of organisms, including the dinosaurs. Flowering plants and the first mammals and birds evolved during this era. The Mesozoic Era ended with another episode of mass extinction, eliminating the dinosaurs and setting the stage for the adaptive radiation of mammals.

The present Cenozoic Era has been marked by the evolution of all of the modern orders of mammals and the explosive radiation of flowering plants. In the past 2.5 million years, a series of intermittent ice ages has caused mass extinctions and reshaped the distribution of plants and animals across the earth.

Key Terms

chemical evolution (p. 760)
Proterozoic Era (p. 764)
cyanobacteria (p. 764)
Paleozoic Era (p. 767)
continental drift (p. 767)
Mesozoic Era (p. 771)
Cenozoic Era (p. 774)

Review Questions

1. Retrace the presumed steps of chemical evolution that led to the appearance of the first living cell.
2. How have the sources of energy important for biological evolution changed from the time the earth formed to the present day?
3. Which of the four major geologic eras saw the appearance of the first living cells? The first eukaryotes? The first protists? The first gymnosperms? The first flowering plants? The first vertebrates? The first reptiles? The first mammals? The first humans?
4. How has the distribution of land masses changed over the past 200 million years?

Critical Thinking Questions

1. What characteristics qualify a structure as a precell? Why are these characteristics so important as steps in the formation of a living cell?
2. Thus far, no signs of life, past or present, have been found on Venus or Mars. Research the environmental conditions on these planets and develop an explanation for why this is so and why the environment on earth is more hospitable to life.
3. According to the second law of thermodynamics, the universe continues to increase toward a state of increasing disorder. How is it that events could take place on earth that led to the evolution of increasingly more complex structures—from simple molecules to organic compounds to simple prokaryotic cells to eukaryotic cells and, finally, to multicellular plants and animals of increasingly complex structures?
4. Develop a graphic illustration *to scale* for the geologic time line, showing the events listed below: origin of earth; origin of first prokaryotic cells; emergence of first autotrophs; rise of the first eukaryotic cells; appearance of the first multicellular organisms; appearance of all the major animal phyla; the dinosaur era; beginning of the age of mammals; appearance of first primates; evolution of Australopithecines; first humans; first appearance of *Homo sapiens.* Indicate the boundaries of the four major geologic eras and the time of the major extinctions.
5. Try to envision the future, 10, 50, and 100 years from now. Where do you think cultural evolution will have led humans? How will humans have affected the evolution *and* extinction of other organisms on earth? What priorities should biologists establish today to circumvent any negative changes in the future?

Additional Readings

Colbert, E. H. 1989. *Digging into the past.* New York: Dembner. (Introductory)

Edey, M. A., and D. C. Johanson. 1989. *Blueprints: Solving the mystery of evolution.* Boston: Little, Brown. (Intermediate)

Hively, W. 1993. Life beyond boiling. *Discover* May:86–91. (Intermediate)

Horgan, J. 1991. In the beginning. *Sci. Amer.* Feb:116–125. (Intermediate)

Knoll, A. H. 1991. End of the Proterozoic Era. *Sci. Amer.* Oct:64–73. (Intermediate)

Paul, G. S. 1989. Giant meteor impacts and great eruptions: Dinosaur killers? *Bioscience.* March:162–172. (Intermediate)

Wilson, E. O. 1992. *The diversity of life.* Cambridge, MA: Harvard University Press. (Intermediate) (If you read only one other book on biology, consider this one.)

Woese, C. R. 1984. *The origin of life.* Carolina Biological Supply Co., Burlington, North Carolina. (Intermediate)

York, D. 1993. The earliest history of the earth. *Sci. Amer.* Jan:90–96. (Intermediate)

APPENDIX A

Metric and Temperature Conversion Charts

Metric Unit (symbol)	*Metric to English*	*English to Metric*
Length		
kilometer (km) = 1,000 (10^3) meters	1 km = 0.62 mile	1 mile = 1.609 km
meter (m) = 100 centimeters	1 m = 1.09 yards = 3.28 feet	1 yard = 0.914 m 1 foot = 0.305 m
centimeter (cm) = 0.01 (10^{-2}) meter	1 cm = 0.394 inch	1 inch = 2.54 cm
millimeter (mm) = 0.001 (10^{-3}) meter	1 mm = 0.039 inch	1 inch = 25.4 mm
micrometer (μm) = 0.000001 (10^{-6}) meter		
nanometer (nm) = 0.000000001 (10^{-9}) meter		
angstrom (Å) = 0.0000000001 (10^{-10}) meter		
Area		
square kilometer (km^2) = 100 hectares	1 km^2 = 0.386 square mile	1 square mile = 2.590 km^2
hectare (ha) = 10,000 square meters	1 ha = 2.471 acres	1 acre = 0.405 ha
square meter (m^2) = 10,000 square centimeters	1 m^2 = 1.196 square yards = 10.764 square feet	1 square yard = 0.836 m^2 1 square foot = 0.093 m^2
square centimeter (cm^2) = 100 square millimeters	1 cm^2 = 0.155 square inch	1 square inch = 6.452 cm^2
Mass		
metric ton (t) = 1,000 kilograms = 1,000,000 grams	1 t = 1.103 tons	1 ton = 0.907 t
kilogram (kg) = 1,000 grams	1 kg = 2.205 pounds	1 pound = 0.454 kg
gram (g) = 1,000 milligrams	1 g = 0.035 ounce	1 ounce = 28.35 g
milligram (mg) = 0.001 gram		
microgram (μg) = 0.000001 gram		
Volume Solids		
1 cubic meter (m^3) = 1,000,000 cubic centimeters	1 m^3 = 1.308 cubic yards = 35.315 cubic feet	1 cubic yard = 0.765 m^3 1 cubic foot = 0.028 m^3
1 cubic centimeter (cm^3) = 1,000 cubic millimeters	1 cm^3 = 0.061 cubic inch	1 cubic inch = 16.387 cm^3
Volume Liquids		
kiloliter (kl) = 1,000 liters	1 kl = 264.17 gallons	
liter (l) = 1,000 milliliters	1 l = 1.06 quarts	1 gal = 3.785 l 1 qt = 0.94 l 1 pt = 0.47 l
milliliter (ml) = 0.001 liter	1 ml = 0.034 fluid ounce	1 fluid ounce = 29.57 ml
microliter (μl) = 0.000001 liter		

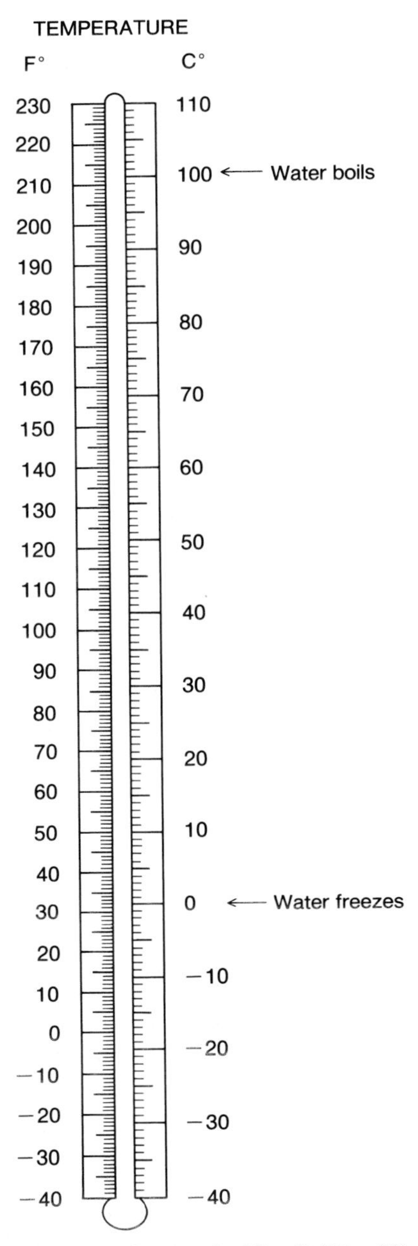

Fahrenheit to Centigrade: $°C = \frac{5}{9}(°F - 32)$

Centigrade to Fahrenheit: $°F = \frac{9}{5}(°C + 32)$

APPENDIX B

Microscopes: Exploring the Details of Life

Microscopes are the instruments that have allowed biologists to visualize objects that are vastly smaller than anything visible with the naked eye. There are broadly two types of specimens viewed in a microscope: whole mounts which consist of an intact subject, such as a hair, a living cell, or even a DNA molecule, and thin sections of a specimen, such as a cell or piece of tissue.

THE LIGHT MICROSCOPE

A light microscope consists of a series of glass lenses that bend (refract) the light coming from an illuminated specimen so as to form a visual image of the specimen that is larger than the specimen itself (a). The specimen is often stained with a colored dye to increase its visibility. A special phase contrast light microscope is best suited for observing unstained, living cells because it converts differences in the density of cell organelles, which are normally invisible to the eye, into differences in light intensity which can be seen.

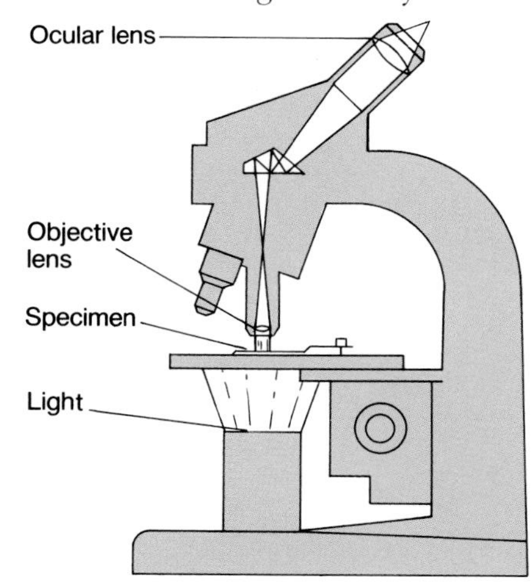

(a)

All light microscopes have limited *resolving power*—the ability to distinguish two very close objects as being separate from each other. The resolving power of the light microscope is about 0.2 μm (about 1,000 times that of the naked eye), a property determined by the wave length of visible light. Consequently, objects closer to each other than 0.2 μm, which includes many of the smaller cell organelles, will be seen as a single, blurred object through a light microscope.

THE TRANSMISSION ELECTRON MICROSCOPE

Appreciation of the wondrous complexity of cellular organization awaited the development of the transmission electron microscope (or TEM), which can deliver resolving powers 1000 times greater than the light microscope. Suddenly, biologists could see strange new structures, whose function was totally unknown—a breakthrough that has kept cell biologists busy for the past 50 years. The TEM (b) works by shooting a beam of electrons through very thinly sliced specimens that have been stained with heavy metals,

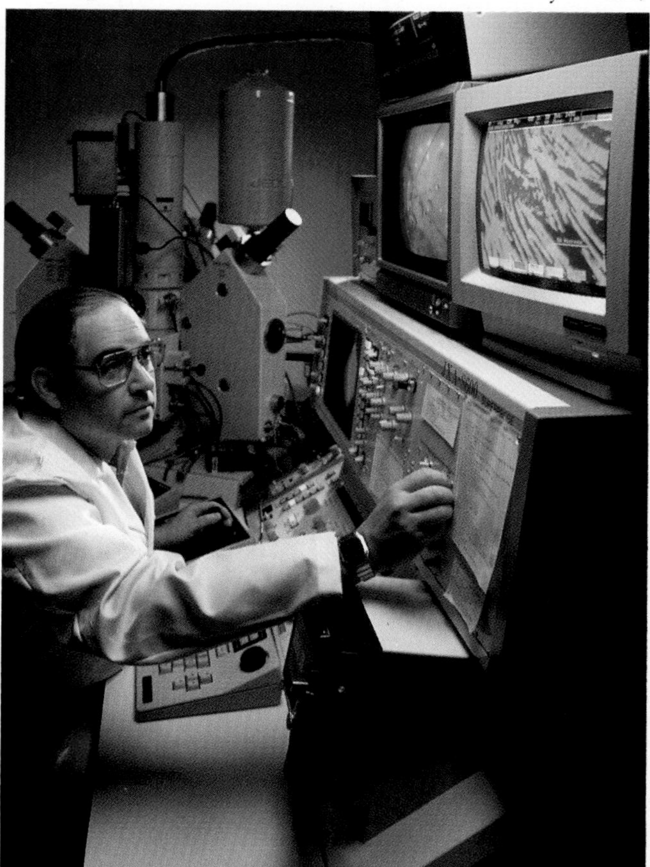
(b)

such as uranium, capable of deflecting electrons in the beam. The electrons that pass through the specimen undeflected are focused by powerful electromagnets (the lenses of a TEM) onto either a phosphorescent screen or high-contrast photographic film. The resolution of the TEM is so great—sufficient to allow us to see individual DNA molecules—because the wavelength of an electron beam is so small (about 0.0005 μm).

THE SCANNING ELECTRON MICROSCOPE

Specimens examined in the scanning electron microscope (SEM) are whole mounts whose surfaces have been coated with a thin layer of heavy metals. In the SEM, a fine beam of electrons scans back and forth across the specimen and the image is formed from electrons bouncing off the hills and valleys of its surface. The SEM produces a three-dimensional image of the surface of the specimen—which can

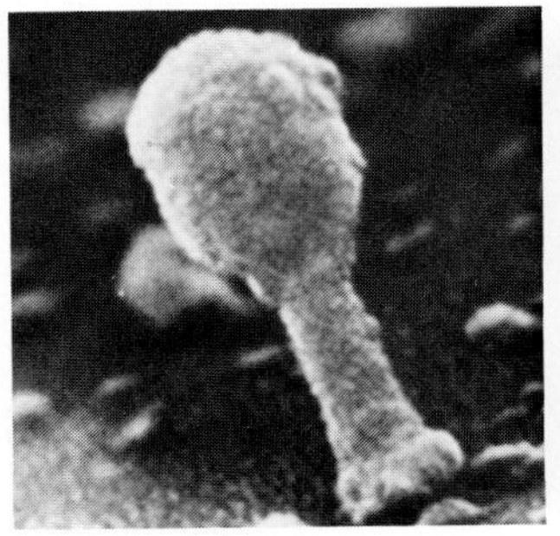

(c)

(d)

range in size from a virus to an insect head (c,d)—with remarkable depth and clarity. The SEM produces black and white images; the colors seen in many of the micrographs in the text have been added to enhance their visual quality. Note that the insect head (d) is that of an antennatedia mutant as described on p 687.

APPENDIX C

The Hardy-Weinberg Principle

If the allele for brown hair is dominant over that for blond hair, and curly hair is dominant over straight hair, then why don't all people by now have brown, curly hair? The **Hardy-Weinberg Principle** (developed independently by English mathematician G. H. Hardy and German physician W. Weinberg) demonstrates that the frequency of alleles remains the same from generation to generation unless influenced by outside factors. The outside factors that would cause allele frequencies to change are mutation, immigration and emigration (movement of individuals into and out of a breeding population, respectively), natural selection of particular traits, and breeding between members of a small population. In other words, unless one or more of these forces influence hair color and hair curl, the relative number of people with brown and curly hair will not increase over those with blond and straight hair.

To illustrate the Hardy-Weinberg Principle, consider a single gene locus with two alleles, *A* and *a*, in a breeding population. (If you wish, consider *A* to be the allele for brown hair and *a* to be the allele for blond hair.) Because there are only two alleles for the gene, the sum of the frequencies of *A* and *a* will equal 1.0. (By convention, allele frequencies are given in decimals instead of percentages.) Translating this into mathematical terms, if

p = the frequency of allele A, and
q = the frequency of allele a,

then $p + q = 1$.

If A represented 80 percent of the alleles in the breeding population ($p = 0.8$), then according to this formula the frequency of a must be 0.2 ($p + q = 0.8 + 0.2 = 1.0$).

After determining the allele frequency in a starting population, the predicted frequencies of alleles and genotypes in the next generation can be calculated. Setting up a Punnett square with starting allele frequencies of $p = 0.8$ and $q = 0.2$:

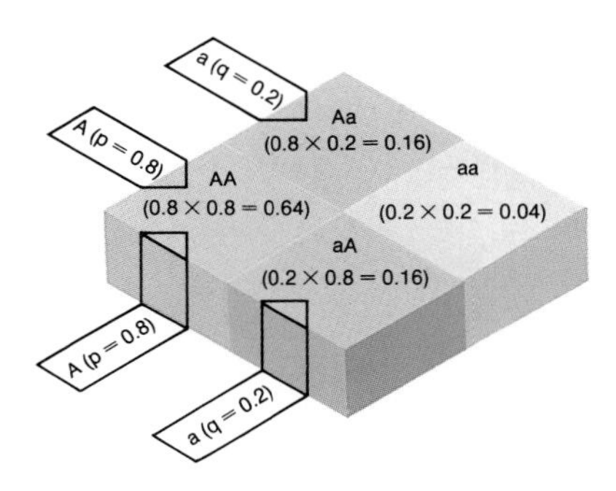

The chances of each offspring receiving any combination of the two alleles is the product of the probability of receiving one of the two alleles alone. In this example, the chances of an offspring receiving two A alleles is $p \times p = p^2$, or $0.8 \times 0.8 = 0.64$. A frequency of 0.64 means that 64 percent of the next generation will be homozygous dominant (*AA*). The chances of an offspring receiving two *a* alleles is $q^2 = 0.2 \times 0.2 = 0.04$, meaning 4 percent of the next generation is predicted to be *aa*. The predicted frequency of heterozygotes (*Aa* or *aA*) is 0.32 or $2pq$, the sum of the probability of an individual being $Aa(p \times q = 0.8 \times 0.2 = 0.16)$ plus the probability of an individual being $aA(q \times p = 0.2 \times 0.8 = 0.16)$. Just as all of the allele frequencies for a particular gene must add up to 1, so must all of the possible genotypes for a particular gene locus add up to 1. Thus, the Hardy-Weinberg Principle is

$$p^2 + 2pq + q^2 = 1$$
$$(0.64 + 0.32 + 0.04 = 1)$$

So after one generation, the frequency of possible genotypes is

$$AA = p^2 = 0.64$$
$$Aa = 2pq = 0.32$$
$$aa = q^2 = 0.04$$

Now let's determine the actual allele frequencies for A and *a* in the new generation. (Remember the original allele frequencies were 0.8 for allele A and 0.2 for allele *a*. If the Hardy-Weinberg Principle is right, there will be no change in the frequency of either allele.) To do this we sum the frequencies for each genotype containing the allele. Since heterozygotes carry both alleles, the genotype frequency must be divided in half to determine the frequency of each allele. (In our example, heterozygote *Aa* has a frequency of 0.32, 0.16 for allele A, plus 0.16 for allele *a*.) Summarizing then:

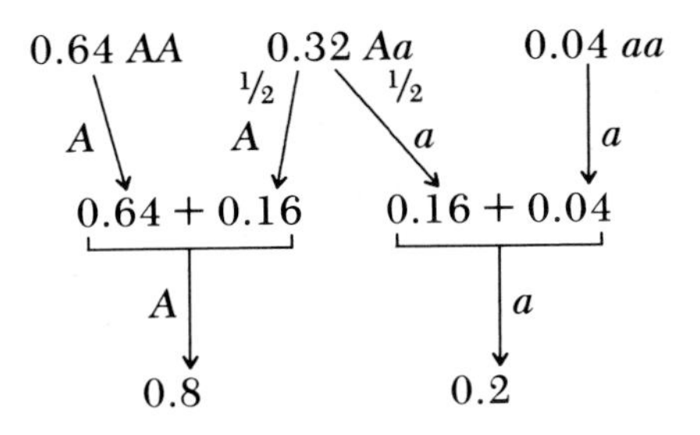

As predicted by the Hardy-Weinberg Principle, the frequency of allele A remained 0.8 and the frequency of allele *a* remained 0.2 in the new generation. Future generations can be calculated in exactly the same way, over and over again. As long as there are no mutations, no gene flow between populations, completely random mating, no natural selection, and no genetic drift, there will be no change in allele frequency, and therefore no evolution.

Population geneticists use the Hardy-Weinberg Principle to calculate a starting point allele frequency, a reference that can be compared to frequencies measured at some future time. The amount of deviation between observed allele frequencies and those predicted by the Hardy-Weinberg Principle indicates the degree of evolutionary change. Thus, this principle enables population geneticists to measure the rate of evolutionary change and identify the forces that cause changes in allele frequency.

APPENDIX D

Careers in Biology

Although many of you are enrolled in biology as a requirement for another major, some of you will become interested enough to investigate the career opportunities in life sciences. This interest in biology can grow into a satisfying livelihood. Here are some facts to consider:

- Biology is a field that offers a very wide range of possible science careers
- Biology offers high job security since many aspects of it deal with the most vital human needs: health and food
- Each year in the United States, nearly 40,000 people obtain bachelor's degrees in biology. But the number of newly created and vacated positions for biologists is increasing at a rate that exceeds the number of new graduates. Many of these jobs will be in the newer areas of biotechnology and bioservices.

Biologists not only enjoy job satisfaction, their work often changes the future for the better. Careers in medical biology help combat diseases and promote health. Biologists have been instrumental in preserving the earth's life-supporting capacity. Biotechnologists are engineering organisms that promise dramatic breakthroughs in medicine, food production, pest management, and environmental protection. Even the economic vitality of modern society will be increasingly linked to biology.

Biology also combines well with other fields of expertise. There is an increasing demand for people with backgrounds or majors in biology complexed with such areas as business, art, law, or engineering. Such a distinct blend of expertise gives a person a special advantage.

The average starting salary for all biologists with a Bachelor's degree is $22,000. A recent survey of California State University graduates in biology revealed that most were earning salaries between $20,000 and $50,000. But as important as salary is, most biologists stress job satisfaction, job security, work with sophisticated tools and scientific equipment, travel opportunities (either to the field or to scientific conferences), and opportunities to be creative in their job as the reasons they are happy in their career.

Here is a list of just a few of the careers for people with degrees in biology. For more resources, such as lists of current openings, career guides, and job banks, write to Biology Career Information, John Wiley and Sons, 605 Third Avenue, New York, NY 10158.

A SAMPLER OF JOBS THAT GRADUATES HAVE SECURED IN THE FIELD OF BIOLOGY*

Agricultural Biologist
Agricultural Economist
Agricultural Extension Officer
Agronomist
Amino-acid Analyst
Analytical Biochemist
Anatomist
Animal Behavior Specialist
Anticancer Drug Research Technician
Antiviral Therapist
Arid Soils Technician
Audio-neurobiologist
Author, Magazines & Books
Behavioral Biologist
Bioanalyst
Bioanalytical Chemist
Biochemical/Endocrine Toxicologist
Biochemical Engineer
Pharmacology Distributor
Pharmacology Technician
Biochemist
Biogeochemist
Biogeographer
Biological Engineer
Biologist
Biomedical Communication Biologist
Biometerologist
Biophysicist
Biotechnologist
Blood Analyst
Botanist
Brain Function Researcher
Cancer Biologist
Cardiovascular Biologist
Cardiovascular/Computer Specialist
Chemical Ecologist
Chromatographer
Clinical Pharmacologist
Coagulation Biochemist
Cognitive Neuroscientist
Computer Scientist
Dental Assistant
Ecological Biochemist
Electrophysiology/Cardiovascular Technician
Energy Regulation Officer
Environmental Biochemist
Environmental Center Director
Environmental Engineer
Environmental Geographer
Environmental Law Specialist
Farmer
Fetal Physiologist
Flavorist
Food Processing Technologist
Food Production Manager
Food Quality Control Inspector
Flower Grower
Forest Ecologist
Forest Economist
Forest Engineer
Forest Geneticist
Forest Manager

Forest Pathologist
Forest Plantation Manager
Forest Products Technologist
Forest Protection Expert
Forest Soils Analyst
Forester
Forestry Information Specialist
Freeze-Dry Engineer
Fresh Water Biologist
Grant Proposal Writer
Health Administrator
Health Inspector
Health Scientist
Hospital Administrator
Hydrologist
Illustrator
Immunochemist
Immunodiagnostic
Assay Developer
Inflammation Technologist
Landscape Architect
Landscape Designer
Legislative Aid
Lepidopterist
Liaison Scientist,
Library of Medicine
Computer Biologist
Life Science Computer
Technologist
Lipid Biochemist
Livestock Inspector
Lumber Inspector
Medical Assistant
Medical Imaging Technician
Medical Officer
Medical Products Developer
Medical Writer
Microbial Physiologist
Microbiologist
Mine Reclamation Scientist
Molecular Endocrinologist
Molecular Neurobiologist
Molecular Parasitologist
Molecular Toxicologist
Molecular Virologist
Morphologist
Natural Products Chemist
Natural Resources Manager
Nature Writer
Nematode Control Biologist
Nematode Specialist
Nematologist
Neuroanatomist
Neurobiologist
Neurophysiologist
Neuroscientist
Nucleic Acids Chemist
Nursing Aid
Nutritionist
Occupational Health Officer
Ornamental Horticulturist
Paleontologist
Paper Chemist
Parasitologist
Pathologist
Peptide Biochemist
Pharmaceutical Writer
Pharmaceutical Sales
Pharmacologist
Physiologist
Planning Consultant
Plant Pathologist
Plant Physiologist
Production Agronomist
Protein Biochemist
Protein Structure & Design
Technician
Purification Biochemist
Quantitative Geneticist
Radiation Biologist
Radiological Scientist
Regional Planner
Regulatory Biologist
Renal Physiologist
Renal Toxicologist
Reproductive Toxicologist
Research and Development
Director
Research Technician
Research Liaison Scientist
Research Products Designer
Research Proposal Writer
Safety Assessment Sanitarian
Scientific Illustrator
Scientific Photographer
Scientific Reference Librarian
Scientific Writer
Soil Microbiologist
Space Station Life Support
Technician
Spectroscopist
Sports Product Designer
Steroid Health Assessor
Taxonomic Biologist
Teacher
Technical Analyst
Technical Science Project
Writer
Textbook Editor
Theoretical Ecologist
Timber Harvester
Toxicologist
Toxic Waste Treatment
Specialist
Urban Planner
Water Chemist
Water Resources Biologist
Wood Chemist
Wood Fuel Technician
Zoning and Planning
Manager
Zoologist
Zoo Animal Breeder
Zoo Animal Behaviorist
Zoo Designer
Zoo Inspector

*Results of one survey of California State University graduates. Some careers may require advanced degrees

Glossary

◀ A ▶

Abiotic Environment Components of ecosystems that include all nonliving factors. (41)

Abscisic Acid (ABA) A plant hormone that inhibits growth and causes stomata to close. ABA may not be commonly involved in leaf drop. (21)

Abscission Separation of leaves, fruit, and flowers from the stem. (21)

Acclimation A physiological adjustment to environmental stress. (28)

Acetyl CoA Acetyl coenzyme A. A complex formed when acetic acid binds to a carrier coenzyme forming a bridge between the end products of glycolysis and the Krebs cycle in respiration. (9)

Acetylcholine Neurotransmitter released by motor neurons at neuromuscular junctions and by some interneurons. (23)

Acid Rain Occurring in polluted air, rain that has a lower pH than rain from areas with unpolluted air. (40)

Acids Substances that release hydrogen ions (H^+) when dissolved in water. (3)

Acid Snow Occurring in polluted air, snow that has a lower pH than snow from areas with unpolluted air. (40)

Acoelomates Animals that lack a body cavity between the digestive cavity and body wall. (39)

Acquired Immune Deficiency Syndrome (AIDS) Disease caused by infection with HIV (Human Immunodeficiency Virus) that destroys the body's ability to mount an immune response due to destruction of its helper T cells. (30, 36)

Actin A contractile protein that makes up the major component of the thin filaments of a muscle cell and the microfilaments of nonmuscle cells. (26)

Action Potential A sudden, dramatic reversal of the voltage (potential difference) across the plasma membrane of a nerve or muscle cell due to the opening of the sodium channels. The basis of a nerve impulse. (23)

Activation Energy Energy required to initiate chemical reaction. (6)

Active Site Region on an enzyme that binds its specific substrates, making them more reactive. (6)

Active Transport Movement of substances into or out of cells against a concentration gradient, i.e., from a region of lower concentration to a region of higher concentration. The process requires an expenditure of energy by the cell. (7)

Adaptation A hereditary trait that improves an organism's chances of survival and/or reproduction. (33)

Adaptive Radiation The divergence of many species from a single ancestral line. (33)

Adenosine Triphosphate (ATP) The molecule present in all living organisms that provides energy for cellular reactions in its phosphate bonds. ATP is the universal energy currency of cells. (6)

Adenylate Cyclase An enzyme activated by hormones that converts ATP to cyclic AMP, a molecule that activates resting enzymes. (25)

Adrenal Cortex Outer layer of the adrenal glands. It secretes steroid hormones in response to ACTH. (25)

Adrenal Medulla An endocrine gland that controls metabolism, cardiovascular function, and stress responses. (25)

Adrenocorticotropic Hormone (ACTH) An anterior pituitary hormone that stimulates the cortex of the adrenal glands to secrete cortisol and other steroid hormones. (25)

Adventitious Root System Secondary roots that develop from stem or leaf tissues. (18)

Aerobe An organism that requires oxygen to release energy from food molecules. (9)

Aerobic Respiration Pathway by which glucose is completely oxidized to CO_2 and H_2O, requiring oxygen and an electron transport system. (9)

Afferent (Sensory) Neurons Neurons that conduct impulses from the sense organs to the central nervous system. (23)

Age-Sex Structure The number of individuals of a certain age and sex within a population. (43)

Aggregate Fruits Fruits that develop from many pistils in a single flower. (20)

AIDS See Acquired Immune Deficiency Syndrome.

Albinism A genetic condition characterized by an absence of epidermal pigmentation that can result from a deficiency of any of a variety of enzymes involved in pigment formation. (12)

Alcoholic Fermentation The process in which electrons removed during glycolysis are transferred from NADH to form alcohol as an end product. Used by yeast during the commercial process of ethyl alcohol production. (9)

Aldosterone A hormone secreted by the adrenal cortex that stimulates reabsorption of sodium from the distal tubules and collecting ducts of the kidneys. (23)

Algae Any unicellular or simple colonial photosynthetic eukaryote. (37,38)

Algin A substance produced by brown algae harvested for human application because of its ability to regulate texture and consistency of products. Found in ice cream, cosmetics, marshmellows, paints, and dozens of other products. (38)

Allantois Extraembryonic membrane that serves as a repository for nitrogenous wastes. In placental mammals, it helps form the vascular connections between mother and fetus. (32)

Allele Alternative form of a gene at a particular site, or locus, on the chromosome. (12)

Allele Frequency The relative occurrence of a certain allele in individuals of a population. (33)

Allelochemicals Chemicals released by some plants and animals that deter or kill a predator or competitor. (42)

Allelopathy A type of interaction in which one organism releases allelochemicals that harm another organism. (42)

Allergy An inappropriate response by the immune system to a harmless foreign substance leading to symptoms such as itchy eyes, runny nose, and congested airways. If the reaction occurs throughout the body (anaphylaxis) it can be life threatening. (30)

Allopatric Speciation Formation of new species when gene flow between parts of a population is stopped by geographic isolation. (33)

Alpha Helix Portion of a polypeptide chain organized into a defined spiral conformation. (4)

Alternation of Generations Sequential change during the life cycle of a plant in which a haploid (1N) multicellular stage (gametophyte) alternates with a diploid (2N) multicellular stage (sporophyte). (38)

Alternative Processing When a primary RNA transcript can be processed to form more than one mRNA depending on conditions. (15)

Altruism The performance of a behavior that benefits another member of the species at some cost to the one who does the deed. (44)

Alveolus A tiny pouch in the lung where gas is exchanged between the blood and the air; the functional unit of the lung where CO_2 and O_2 are exchanged. (29)

Alzheimer's Disease A degenerative disease of the human brain, particularly affecting acetylcholine-releasing neurons and the hippocampus, characterized by the presence of tangled fibrils within the cytoplasm of neurons and amyloid plaques outside the cells. (23)

Amino Acids Molecules containing an amino group ($-NH_2$) and a carboxyl group ($-COOH$) attached to a central carbon atom. Amino acids are the subunits from which proteins are constructed. (4)

Amniocentesis A procedure for obtaining fetal cells by withdrawing a sample of the fluid that surrounds a developing fetus (amniotic fluid) using a hypodermic needle and syringe. (17)

Amnion Extraembryonic membrane that envelops the young embryo and encloses the amniotic fluid that suspends and cushions it. (32)

Amoeba A protozoan that employs pseudopods for motility. (37)

Amphibia A vertebrate class grouped into three orders: Caudata (tailed amphibians); Anura (tail-less amphibians); Apoda (rare wormlike, burrowing amphibians). (39)

Anabolic Steroids Steroid hormones, such as testosterone, which promote biosynthesis (anabolism), especially protein synthesis. (25)

Anabolism Biosynthesis of complex molecules from simpler compounds. Anabolic pathways are endergonic, i.e., require energy. (6)

Anaerobe Organism that does not require oxygen to release energy from food molecules. (9)

Anaerobic Respiration Pathway by which glucose is completely oxidized, using an electron transport system but requiring a terminal electron acceptor other than oxygen. (Compare with fermentation.) (9)

Analogous Structures (Homoplasies) Structures that perform a similar function, such as wings in birds and insects, but did not originate from the same structure in a common ancestor. (33)

Anaphase Stage of mitosis when the kinetochores split and the sister chromatids (now termed chromosomes) move to opposite poles of the spindle. (10)

Anatomy Study of the structural characteristics of an organism. (18)

Angiosperm (Anthophyta) Any plant having its seeds surrounded by fruit tissue formed from the mature ovary of the flowers. (38)

Animal A mobile, heterotrophic, multicellular organism, classified in the Animal kingdom. (39)

Anion A negatively charged ion. (3)

Annelida The phylum which contains segmented worms (earthworms, leeches, and bristleworms). (39)

Annuals Plants that live for one year or less. (18)

Annulus A row of specialized cells encircling each sporangium on the fern frond; facilitates rupture of the sporangium and dispersal of spors. (38)

Antagonistic Muscles Pairs of muscles whose contraction bring about opposite actions as illustrated by the biceps and triceps, which bends or straightens the arm at the elbow, respectively. (26)

Antenna Pigments Components of photosystems that gather light energy of different wavelengths and then channel the absorbed energy to a reaction center. (8)

Anterior In anatomy, at or near the front of an animal; the opposite of posterior. (39)

Anterior Pituitary A true endocrine gland manufacturing and releasing six hormones when stimulated by releasing factors from the hypothalamus. (25)

Anther The swollen end of the stamen (male reproductive organ) of a flowering plant. Pollen grains are produced inside the anther lobes in pollen sacs. (20)

Antibiotic A substance produced by a fungus or bacterium that is capable of preventing the growth of bacteria. (2)

Antibodies Proteins produced by plasma cells. They react specifically with the antigen that stimulated their formation. (30)

Anticodon Triplet of nucleotides in tRNA that recognizes and base pairs with a particular codon in mRNA. (14)

Antidiuretic Hormone (ADH) One of the two hormones released by the posterior pituitary. ADH increases water reabsorption in the kidney, which then produces a more concentrated urine. (25)

Antigen Specific foreign agent that triggers an immune response. (30)

Aorta Largest blood vessel in the body through which blood leaves the heart and enters the systemic circulation. (28)

Apical Dominance The growth pattern in plants in which axillary bud growth is inhibited by the hormone auxin, present in high concentrations in terminal buds. (21)

Apical Meristems Centers of growth located at the tips of shoots, axillary buds, and roots. Their cells divide by mitosis to produce new cells for primary growth in plants. (18)

Aposematic Coloring Warning coloration which makes an organism stand out from its surroundings. (42)

Appendicular Skeleton The bones of the appendages and of the pectoral and pelvic girdles. (26)

Aquatic Living in water. (40)

Archaebacteria Members of the kingdom Monera that differ from typical bacteria in the structure of their membrane lipids, their cell walls, and some characteristics that resemble those of eukaryotes. Their lack of a true nucleus, however, accounts for their assignment to the Moneran kingdom. (36)

Archenteron In gastrulation, the hollow core of the gastrula that becomes an animal's digestive tract. (32)

Arteries Large, thick-walled vessels that carry blood away from the heart. (28)

Arterioles The smallest arteries, which carry blood toward capillary beds. (28)

Arthropoda The most diverse phylum on earth, so called from the presence of jointed limbs. Includes insects, crabs, spiders, centipedes. (39)

Ascospores Sexual fungal spore borne in a sac. Produced by the sac fungi, Ascomycota. (37)

Asexual Reproduction Reproduction without the union of male and female gametes. (31)

Association In ecological communities, a major organization characterized by uniformity and two or more dominant species. (41)

Asymmetric Referring to a body form that cannot be divided to produce mirror images. (39)

Atherosclerosis Condition in which the inner walls of arteries contain a buildup of cholesterol-containing plaque that tends to occlude the channel and act as a site for the formation of a blood clot (thrombus). (7)

Atmosphere The layer of air surrounding the Earth. (40)

Atom The fundamental unit of matter that can enter into chemical reactions; the smallest unit of matter that possesses the qualities of an element. (3)

Atomic Mass Combined number of protons and neutrons in the nucleus of an atom. (3)

Atomic Number The number of protons in the nucleus of an atom. (3)

ATP (see **Adenosine Triphosphate**)

ATPase An enzyme that catalyzes a reaction in which ATP is hydrolyzed. These enzymes are typically involved in reactions where energy stored in ATP is used to drive an energy-requiring reaction, such as active transport or muscle contractility. (7, 26)

ATP Synthase A large protein complex present in the plasma membrane of bacteria, the inner membrane of mitochondria, and the thylakoid membrane of chloroplasts. This complex consists of a baseplate in the membrane, a channel across the membrane through which protons can pass, and a spherical head (F_1 particle) which contains the site where ATP is synthesized from ADP and P_i. (8, 9)

Atrioventricular (AV) Node A neurological center of the heart, located at the top of the ventricles. (28)

Atrium A contracting chamber of the heart which forces blood into the ventricle. There are two atria in the hearts of all vertebrates, except fish which have one atrium. (28)

Atrophy The shrinkage in size of structure, such as a bone or muscle, usually as a result of disuse. (26)

Autoantibodies Antibodies produced against the body's own tissue. (30)

Autoimmune Disease Damage to a body tissue due to an attack by autoantibodies. Examples include thyroiditis, multiple sclerosis, and rheumatic fever. (30)

Autonomic Nervous System The nerves that control the involuntary activities of the internal organs. It is composed of the parasympathetic system, which functions during normal activity, and the sympathetic system, which operates in times of emergency or prolonged exertion. (23)

Autosome Any chromosome that is not a sex chromosome. (13)

Autotrophs Organisms that satisfy their own nutritional needs by building organic molecules photosynthetically or chemosynthetically from inorganic substances. (8)

Auxins Plant growth hormones that promote cell elongation by softening cell walls. (21)

Axial Skeleton The bones aligned along the long axis of the body, including the skull, vertebral column, and ribcage. (26)

Axillary Bud A bud that is directly above each leaf on the stem. It can develop into a new stem or a flower. (18)

Axon The long, sometimes branched extension of a neuron which conducts impulses from the cell body to the synaptic knobs. (23)

◄ B ►

Bacteriophage A virus attacking specific bacteria that multiplies in the bacterial host cell and usually destroys the bacterium as it reproduces. (36)

Balanced Polymorphism The maintenance of two or more alleles for a single trait at fairly high frequencies. (33)

Bark Common term for the periderm. A collective term for all plant tissues outside the secondary xylem. (18)

Base Substance that removes hydrogen ions (H^+) from solutions. (3)

Basidiospores Sexual spores produced by basidiomycete fungi. Often found by the millions on gills in mushrooms. (37)

Basophil A phagocytic leukocyte which also releases substances, such as histamine, that trigger an inflammatory response. (28)

Batesian Mimicry The resemblance of a good-tasting or harmless species to a species with unpleasant traits. (42)

Bathypelagic Zone The ocean zone beneath the mesopelagic zone, characterized by no light; inhabited by heterotrophic bacteria and benthic scavengers. (40)

B Cell A lymphocyte that becomes a plasma cell and produces antibodies when stimulated by an antigen. (30)

Benthic Zone The deepest ocean zone; the ocean floor, inhabitated by bottom dwelling organisms. (40)

Bicarbonate Ion $HCO_{-3.}$ (3, 29)

Biennials Plants that live for two years. (18)

Bilateral Symmetry The quality possessed by organisms whose body can be divided into mirror images by only one median plane. (39)

Bile Salts Detergentlike molecules produced by the liver and stored by the gallbladder that function in lipid emulsification in the small intestine. (27)

Binomial A term meaning "two names" or "two words". Applied to the system of nomenclature for categorizing living things with a genus and species name that is unique for each type of organism. (1)

Biochemicals Organic molecules produced by living cells. (4)

Bioconcentration The ability of an organism to accumulate substances within its' body or specific cells. (41)

Biodiversity Biological diversity of species, including species diversity, genetic diversity, and ecological diversity. (43)

Biogeochemical Cycles The exchanging of chemical elements between organisms and the abiotic environment. (41)

Biological Control Pest control through the use of naturally occurring organisms such as predators, parasites, bacteria, and viruses. (41)

Biological Magnification An increase in concentration of slowly degradable chemicals in organisms at successively higher trophic levels; for example, DDT or PCB's. (41)

Bioluminescence The capability of certain organisms to utilize chemical energy to produce light in a reaction catalyzed by the enzyme luciferase. (9)

Biomass The weight of organic material present in an ecosystem at any one time. (41)

Biome Broad geographic region with a characteristic array of organisms. (40)

Biosphere Zone of the earth's soil, water, and air in which living organisms are found. (40)

Biosynthesis Construction of molecular components in the growing cell and the replacement of these compounds as they deteriorate. (6)

Biotechnology A new field of genetic engineering; more generally, any practical application of biological knowledge. (16)

Biotic Environment Living components of the environment. (40)

Biotic Potential The innate capacity of a population to increase tremendously in size were it not for curbs on growth; maximum population growth rate. (43)

Blade Large, flattened area of a leaf; effective in collecting sunlight for photosynthesis. (18)

Blastocoel The hollow fluid-filled space in a blastula. (32)

Blastocyst Early stage of a mammalian embryo, consisting of a mass of cells enclosed in a hollow ball of cells called the trophoblast. (32)

Blastodisk In bird and reptile development, the stage equivalent to a blastula. Because of the large amount of yolk, cleavage produces two flattened layers of cells with a blastocoel between them. (32)

Blastomeres The cells produced during embryonic cleavage. (32)

Blastopore The opening of the archenteron that is the embryonic predecessor of the anus in vertebrates and some other animals. (32)

Blastula An early developmental stage in many animals. It is a ball of cells that encloses a cavity, the blastocoel. (32)

Blood A type of connective tissue consisting of red blood cells, white blood cells, platelets, and plasma. (28)

Blood Pressure Positive pressure within the cardiovascular system that propels blood through the vessels. (28)

Blooms are massive growths of algae that occur when conditions are optimal for algae proliferation. (37)

Body Plan The general layout of a plant's or animal's major body parts. (39)

Bohr Effect Increased release of O_2 from hemoglobin molecules at lower pH. (29)

Bone A tissue composed of collagen fibers, calcium, and phosphate that serves as a means of support, a reserve of calcium and phosphate, and an attachment site for muscles. (26)

Botany Branch of biology that studies the life cycles, structure, growth, and classification of plants. (18)

Bottleneck A situation in which the size of a species' population drops to a very small number of individuals, which has a major impact on the likelihood of the population recovering its earlier genetic diversity. As occurred in the cheetah population. (33)

Bowman's Capsule A double-layered container that is an invagination of the proximal end of the renal tubule that collects molecules and wastes from the blood. (28)

Brain Mass of nerve tissue composing the main part of the central nervous system. (23)

Brainstem The central core of the brain, which coordinates the automatic, involuntary body processes. (23)

Bronchi The two divisions of the trachea through which air enters each of the two lungs. (29)

Bronchioles The smallest tubules of the respiratory tract that lead into the alveoli of the lungs where gas exchange occurs. (29)

Bryophyta Division of non-vascular terrestrial plants that include liverworts, mosses, and hornworts. (38)

Budding Asexual process by which offspring develop as an outgrowth of a parent. (39)

Buffers Chemicals that couple with free hydrogen and hydroxide ions thereby resisting changes in pH. (3)

Bundle Sheath Parenchyma cells that surround a leaf vein which regulate the uptake and release of materials between the vascular tissue and the mesophyll cells. (18)

C_3 Synthesis The most common pathway for fixing CO_2 in the synthesis reactions of photosynthesis. It is so named because the first detectable organic molecule into which CO_2 is incorporated is a 3-carbon molecule, phosphoglycerate (PGA). (8)

C_4 Synthesis Pathway for fixing CO_2 during the light-independent reactions of photosynthesis. It is so named because the first detectable organic molecule into which CO_2 is incorporated is a 4-carbon molecule. (8)

Calcitonin A thyroid hormone which regulates blood calcium levels by inhibiting its release from bone. (25)

Calorie Energy (heat) necessary to elevate the temperature of one gram of water by one degree Centigrade (1° C). (6)

Calvin Cycle The cyclical pathway in which CO_2 is incorporated into carbohydrate. See C_3 synthesis. (8)

Calyx The outermost whorl of a flower, formed by the sepals. (20)

CAM Crassulacean acid metabolism. A variation of the photosynthetic reactions in plants, biochemically identical to C_4 synthesis except that all reactions occur in the same cell and are separated by time. Because CAM plants open their stomates at night, they have a competitive advantage in hot, dry climates. (8)

Cambium A ring or cluster of meristematic cells that increase the width of stems and roots when they divide to produce secondary tissues. (18)

Camouflage Adaptations of color, shape and behavior that make an organism more difficult to detect. (42)

Cancer A disease resulting from uncontrolled cell divisions. (10,13)

Capillaries The tiniest blood vessels consisting of a single layer of flattened cells. (28)

Capillary Action Tendency of water to be pulled into a small-diameter tube. (3)

Carbohydrates A group of compounds that includes simple sugars and all larger molecules constructed of sugar subunits, e.g. polysaccharides. (4)

Carbon Cycle The cycling of carbon in different chemical forms, from the environment to organisms and back to the environment. (41)

Carbon Dioxide Fixation In photosynthesis, the combination of CO_2 with carbon-accepting molecules to form organic compounds. (8)

Carcinogen A cancer-causing agent. (13)

Cardiac Muscle One of the three types of muscle tissue; it forms the muscle of the heart. (26)

Cardiovascular System The organ system consisting of the heart and the vessels through which blood flows. (28)

Carnivore An animal that feeds exclusively on other animals. (42)

Carotenoid A red, yellow, or orange plant pigment that absorbs light in 400-500 nm wavelengths. (8)

Carpels Central whorl of a flower containing the female reproductive organs. Each separate carpel, or each unit of fused carpels, is called a pistil. (20)

Carrier Proteins Proteins within the plasma membrane that bind specific substances and facilitate their movement across the membrane. (7)

Carrying Capacity The size of a population that can be supported indefinitely in a given environment. (43)

Cartilage A firm but flexible connective tissue. In the human, most cartilage originally present in the embryo is transformed into bones. (26)

Casparian Strip The band of waxy suberin that surrounds each endodermal cell of a plant's root tissue. (18)

Catabolism Metabolic pathways that degrade complex compounds into simpler molecules, usually with the release of the chemical energy that held the atoms of the larger molecule together. (6)

Catalyst A chemical substance that accelerates a reaction or causes a reaction to occur but remains unchanged by the reaction. Enzymes are biological catalysts. (6)

Cation A positively charged ion. (3)

Cecum A closed-ended sac extending from the intestine in grazing animals lacking a rumen (e.g., horses) that enables them to digest cellulose. (27)

Cell The basic structural unit of all organisms. (5)

Cell Body Region of a neuron that contains most of the cytoplasm, the nucleus, and other organelles. It relays impulses from the dendrites to the axon. (23)

Cell Cycle Complete sequence of stages from one cell division to the next. The stages are denoted G_1, S, G_2, and M phase. (10)

Cell Differentiation The process by which the internal contents of a cell become assembled into a structure that allows the cell to carry out a specific set of activities, such as secretion of enzymes or contraction. (32)

Cell Division The process by which one cell divides into two. (10)

Cell Fusion Technique whereby cells are caused to fuse with one another producing a large cell with a common cytoplasm and plasma membrane. (5, 10)

Cell Plate In plants, the cell wall material deposited midway between the daughter cells during cytokinesis. Plate material is deposited by small Golgi vesicles. (5, 10)

Cell Sap Solution that fills a plant vacuole. In addition to water, it may contain pigments, salts, and even toxic chemicals. (5)

Cell Theory The fundamental theory of biology that states: 1) all organisms are composed of one or more cells, 2) the cell is the basic organizational unit of life, 3) all cells arise from pre-existing cells. (5)

Cellular Respiration (See **Aerobic respiration**)

Cellulose The structural polysaccharide comprising the bulk of the plant cell wall. It is the most abundant polysaccharide in nature. (4, 5)

Cell Wall Rigid outer-casing of cells in plants and other organisms which gives support, slows dehydration, and prevents a cell from bursting when internal pressure builds due to an influx of water. (5)

Central Nervous System In vertebrates, the brain and spinal cord. (23)

Centriole A pinwheel-shaped structure at each pole of a dividing animal cell. (10)

Centromere Indented region of a mitotic chromosome containing the kinetochore. (10)

Cephalization The clustering of neural tissues and sense organs at the anterior (leading) end of the animal. (39)

Cerebellum A bulbous portion of the vertebrate brain involved in motor coordination. Its prominence varies greatly among different vertebrates . (23)

Cerebral Cortex The outer, highly convoluted layer of the cerebrum. In the human, this is the center of higher brain functions, such as speech and reasoning. (23)

Cerebrospinal Fluid Fluid present within the ventricles of the brain, central canal of the spinal cord, and which surrounds and cushions the central nervous system. (23)

Cerebrum The most dominant part of the human forebrain, composed of two cerebral hemispheres, generally associated with higher brain functions. (23)

Cervix The lower tip of the uterus. (31)

Chapparal A type of shrubland in California, characterized by drought-tolerant and fire-adapted plants. (40)

Character Displacement Divergence of a physical trait in closely related species in response to competition. (42)

Chemical Bonds Linkage between atoms as a result of electrons being shared or donated. (3)

Chemical Evolution Spontaneous synthesis of increasingly complex organic compounds from simpler molecules. (35)

Chemical Reaction Interaction between chemical reactants. (6)

Chemiosmosis The process by which a pH gradient drives the formation of ATP. (8, 9)

Chemoreceptors Sensory receptors that respond to the presence of specific chemicals. (24)

Chemosynthesis An energy conversion process in which inorganic substances (H, N, Fe, or S) provide energized electrons and hydrogen for carbohydrate formation (9, 36)

Chiasmata Cross-shaped regions within a tetrad, occurring at points of crossing over or genetic exchange. (11)

Chitin Structural polysaccharide that forms the hard, strong external skeleton of many arthropods and the cell walls of fungi. (4)

Chlamydia Obligate intracellular parasitic bacteria that lack a functional ATP-generating system. (36)

Chlorophyll Pigments Major light-absorbing pigments of photosynthesis. (8)

Chlorophyta Green algae, the largest group of algae; members of this group were very likely the ancestors of the modern plant kingdom. (38)

Chloroplasts An organelle containing chlorophyll found in plant cells in which photosynthesis occurs. (5, 8)

Cholecystokinin (CCK) Hormone secreted by endocrine cells in the wall of the small intestine that stimulates the release of digestive products by the pancreas. (27)

Chondrocytes Living cartilage cells embedded within the protein-polysaccharide matrix they manufacture. (26)

Chordamesoderm In vertebrates, the block of mesoderm that underlies the dorsal ectoderm of the gastrula, induces the formation of the nervous system, and gives rise to the notochord. (32)

Chordate A member of the phylum Chordata possessing a skeletal rod of tissue called a notochord, a dorsal hollow nerve cord, gill slits, and a post-anal tail at some stage of its development. (39)

Chorion The outermost of the four extraembryonic membranes. In placental mammals, it forms the embryonic portion of the placenta. (32)

Chorionic Villus Sampling (CVS) A procedure for obtaining fetal cells by removing a small sample of tissue from the developing placenta of a pregnant woman. (17)

Chromatid Each of the two identical subunits of a replicated chromosome. (10)

Chromatin DNA-protein fibers which, during prophase, condense to form the visible chromosomes. (5, 10)

Chromatography A technique for separating different molecules on the basis of their solubility in a particular solvent. The mixture of substances is spotted on a piece of paper or other material, one end of which is then placed in the solvent. As the solvent moves up the paper by capillary action, each substance in the mixture is carried a particular distance depending on its solubility in the moving solvent. (8)

Chromosomes Dark-staining structures in which the organism's genetic material (DNA) is organized. Each species has a characteristic number of chromosomes. (5, 10)

Chromosome Aberrations Alteration in the structure of a chromosome from the normal state. Includes chromosome deletions, duplications, inversions, and translocations. (13)

Chromosome Puff A site on an insect polytene chromosome where the DNA has unraveled and is being transcribed. (15)

Cilia Short, hairlike structures projecting from the surfaces of some cells. They beat in coordinated ways, are usually found in large numbers, and are densely packed. (5)

Ciliated Mucosa Layer of ciliated epithelial cells lining the respiratory tract. The beating of cilia propels an associated mucous layer and trapped foreign particles. (29)

Circadian Rhythm Behavioral patterns that cycle during approximately 24 hour intervals.

Circulatory System The system that circulates internal fluids throughout an organism to deliver oxygen and nutrients to cells and to remove metabolic wastes. (28)

Class (Taxonomic) A level of the taxonomic hierarchy that groups together members of related orders. (1)

Classical Conditioning A form of learning in which an animal develops a response to a new stimulus by repeatedly associating the new stimulus with a stimulus that normally elicits the response. (44)

Cleavage Successive mitotic divisions in the early embryo. There is no cell growth between divisions. (32)

Cleavage Furow Constriction around the middle of a dividing cell caused by constriction of microfilaments. (10)

Climate The general pattern of average weather conditions over a long period of time in a specific region, including precipitation, temperature, solar radiation, and humidity. (40)

Climax Final or stable community of successional stages, that is more or less in equilibrium with existing environmental conditions for a long period of time. (41)

Climax Community Community that remains essentially the same over long periods of time; final stage of ecological succession. (41)

Clitoris A protrusion at the point where the labia minora merge; rich in sensory neurons and erectile tissue. (31)

Clonal Selection Mechanism The mechanism by which the body can synthesize antibodies specific for the foreign substance (antigen) that stimulated their production. (30)

Clones Offspring identical to the parent, produced by asexual processes. (15)

Closed Circulatory System Circulatory system in which blood travels throughout the body in a continuous network of closed tubes. (Compare with open circulatory system). (28)

Clumped Pattern Distribution of individuals of a population into groups, such as flocks or herds. (43)

Cnidaria A phylum that consists of radial symmetrical animals that have two cell layers. There are three classes: 1) Hydrozoa (hydra), 2) Scyphozoa (jellyfish), 3) Anthozoa (sea anemones, corals). Most are marine forms that live in warm, shallow water. (39)

Cnidocytes Specialized stinging cells found in the members of the phylum Cnidaria. (39)

Coastal Waters Relatively warm, nutrient-rich shallow water extending from the high-tide mark on land to the sloping continental shelf. The greatest concentration of marine life are found in coastal waters. (40)

Coated Pits Indentations at the surfaces of cells that contain a layer of bristly protein (called clathrin) on the inner surface of the plasma membrane. Coated pits are sites where cell receptors become clustered. (7)

Cochlea Organ within the inner ear of mammals involved in sound reception. (24)

Codominance The simultaneous expression of both alleles at a genetic locus in a heterozygous individual. (12)

Codon Linear array of three nucleotides in mRNA. Each triplet specifies a particular amino acid during the process of translation. (14)

Coelomates Animals in which the body cavity is completely lined by mesodermally-derived tissues. (39)

Coenzyme An organic cofactor, typically a vitamin or a substance derived from a vitamin. (6)

Coevolution Evolutionary changes that result from reciprocal interactions between two species, e.g., flowering plants and their insect pollinators. (33)

Cofactor A non-protein component that is linked covalently or noncovalently to an enzyme and is required by the enzyme to catalyze the reaction. Cofactors may be organic molecules (coenzymes) or metals. (6)

Cohesion The tendency of different parts of a substance to hold together because of forces acting between its molecules. (3)

Coitus Sexual union in mammals. (31)

Coleoptile Sheath surrounding the tip of the monocot seedling, protecting the young stem and leaves as they emerge from the soil. (21)

Collagen The most abundant protein in the human body. It is present primarily in the extracellular space of connective tissues such as bone, cartilage, and tendons. (26)

Collenchyma Living plant cells with irregularly thickened primary cell walls. A supportive cell type often found inside the epidermis of stems with primary growth. Angular, lacunar and laminar are different types of collenchyma cells. (18)

Commensalism A form of symbiosis in which one organism benefits from the union while the other member neither gains nor loses. (42)

Community The populations of all species living in a given area. (41)

Compact Bone The solid, hard outer regions of a bone surrounding the honey-combed mass of spongy bone. (26)

Companion Cell Specialized parenchyma cell associated with a sieve-tube member in phloem. (18)

Competition Interaction among organisms that require the same resource. It is of two types: 1) intraspecific (between members of the same species); 2) interspecific (between members of different species). (42)

Competitive Exclusion Principle (Gause's Principle) Competition in which a winner species captures a greater share of resources, increasing its survival and reproductive capacity. The other species is gradually displaced. (42)

Competitive Inhibition Prevention of normal binding of a substrate to its enzyme by the presence of an inhibitory compound that competes with the substrate for the active site on the enzyme. (6)

Complement Blood proteins with which some antibodies combine following attachment to antigen (the surface of microorganisms). The bound complement punches the tiny holes in the plasma membrane of the foreign cell, causing it to burst. (28)

Complementarity The relationship between the two strands of a DNA molecule determined by the base pairing of nucleotides on the two strands of the helix. A nucleotide with guanine on one strand always pairs with a nucleotide having cytosine on the other strand; similarly with adenine and thymine. (14)

Complete Digestive Systems Systems that have a digestive tract with openings at both ends—a mouth for entry and an anus for exit. (27)

Complete Flower A flower containing all four whorls of modified leaves—sepels, petals, stamen, and carpels. (20)

Compound Chemical substances composed of atoms of more than one element. (3)

Compound Leaf A leaf that is divided into leaflets, with two or more leaflets attached to the petiole. (18)

Concentration Gradient Regions in a system of differing concentration representing potential energy, such as exist in a cell and its environment, that cause molecules to move from areas of higher concentration to lower concentration. (7)

Conditioned Reflex A reflex ("automatic") response to a stimulus that would not normally have elicited the response. Conditioned reflexes develop by repeated association of a new stimulus with an old stimulus that normally elicits the response. (44)

Conformation The three-dimensional shape of a molecule as determined by the spatial arrangement of its atoms. (4)

Conformational Change Change in molecular shape (as occurs, for example, in an enzyme as it catalyzes a reaction, or a myosin molecule during contraction). (6)

Conjugation A method of reproduction in single-celled organisms in which two cells link and exchange nuclear material. (11)

Connective Tissues Tissues that protect, support, and hold together the internal organs and other structures of animals. Includes bone, cartilage, tendons, and other tissues, all of which have large amounts of extracellular material. (22)

Consumers Heterotrophs in a biotic environment that feed on other organisms or organic waste. (41)

Continental Drift The continuous shifting of the earth's land masses explained by the theory of plate tectonics. (35)

Continuous Variation An inheritance pattern in which there is graded change between the two extremes in a phenotype (compare with discontinuous variation). (12)

Contraception The prevention of pregnancy. (31)

Contractile Proteins Actin and myosin, the protein filaments that comprise the bulk of the muscle mass. During contraction of skeletal muscle, these filaments form a temporary association and slide past each other, generating the contractile force. (26)

Control (Experimental) A duplicate of the experiment identical in every way except for the one variable being tested. Use of a control is necessary to demonstrate cause and effect. (2)

Convergent Evolution The evolution of similar structures in distantly related organisms in response to similar environments. (33)

Cork Cambium In stems and roots of perennials, a secondary meristem that produces the outer protective layer of the bark. (18)

Coronary Arteries Large arteries that branch immediately from the aorta, providing oxygen-rich blood to the cardiac muscle. (28)

Corpus Callosum A thick cable composed of hundreds of millions of neurons that connect the right and left cerebral hemispheres of the mammalian brain. (23)

Corpus Luteum In the mammalian ovary, the structure that develops from the follicle after release of the egg. It secretes hormones that prepare the uterine endometrium to receive the developing embryo. (31)

Cortex In the stem or root of plants, the region between the epidermis and the vascular tissues. Composed of ground tissue. In animals, the outermost portion of some organs. (18)

Cotyledon The seed leaf of a dicot embryo containing stored nutrients required for the germinated seed to grow and develop, or a food digesting seed leaf in a monocot embryo. (20)

Countercurrent Flow Mechanism for increasing the exchange of substances or heat from one stream of fluid to another by having the two fluids flow in opposite directions. (29)

Covalent Bonds Linkage between two atoms which share the same electrons in their outermost shells. (3)

Cranial Nerves Paired nerves which emerge from the central stalk of the vertebrate brain and innervate the body. Humans have 12 pairs of cranial nerves. (23)

Cranium The bony casing which surrounds and protects the vertebrate brain. (23)

Cristae The convolutions of the inner membrane of the mitochondrion. Embedded within them are the components of the electron transport system and proton channels for chemiosmosis. (9)

Crossing Over During synapsis, the process by which homologues exchange segments with each other. (11)

Cryptic Coloration A form of camouflage wherein an organism's color or patterning helps it resemble its background. (42)

Cutaneous Respiration The uptake of oxygen across virtually the entire outer body surface. (29)

Cuticle 1) Waxy layer covering the outer cell walls of plant epidermal cells. It retards water vapor loss and helps prevent dehydration. (18) 2) Outer protective, nonliving covering of some animals, such as the exoskeleton of anthropods. (26, 39)

Cyanobacteria A type of prokaryote capable of photosynthesis using water as a source of electrons. Cyanobacteria were responsible for initially creating an O_2-containing atmosphere on earth. (35, 36)

Cyclic AMP (Cyclic adenosine monophosphate) A ring-shaped molecular version of an ATP minus two phosphates. A regulatory molecule formed by the enzyme adenylate cyclase which converts ATP to cAMP. A second messenger. (25)

Cyclic Pathways Metabolic pathways in which the intermediates of the reaction are regenerated while assisting the conversion of the substrate to product. (9)

Cyclic Photophosphorylation A pathway that produces ATP, but not NADPH, in the light reactions of photosynthesis. Energized electrons are shuttled from a reaction center, along a molecular pathway, back to the original reaction center, generating ATP en route. (8)

Cysts Protective, dormant structure formed by some protozoa. (37)

Cytochrome Oxidase A complex of proteins that serves as the final electron carrier in the mitochondrial electron transport system, transferring its electrons to O_2 to form water. (9)

Cytokinesis Final event in eukaryotic cell division in which the cell's cytoplasm and the new nuclei are partitioned into separate daughter cells. (10)

Cytokinins Growth-producing plant hormones which stimulate rapid cell division. (21)

Cytoplasm General term that includes all parts of the cell, except the plasma membrane and the nucleus. (5)

Cytoskeleton Interconnecting network of microfilaments, microtubules, and intermediate filaments that serves as a cell scaffold and provides the machinery for intracellular movements and cell motility. (5)

Cytotoxic (Killer) T Cells A class of T cells capable of recognizing and destroying foreign or infected cells. (30)

◄ D ►

Day Neutral Plants Plants that flower at any time of the year, independent of the relative lengths of daylight and darkness. (21)

Deciduous Trees or shrubs that shed their leaves in a particular season, usually autumn, before entering a period of dormancy. (40)

Deciduous Forest Forests characterized by trees that drop their leaves during unfavorable conditions, and leaf out during warm, wet seasons. Less dense than tropical rain forests. (40)

Decomposers (Saprophytes) Organisms that obtain nutrients by breaking down organic compounds in wastes and dead organisms. Includes fungi, bacteria, and some insects. (41)

Deletion Loss of a portion of a chromosome, following breakage of DNA. (13)

Denaturation Change in the normal folding of a protein as a result of heat, acidity, or alkalinity. Such changes result in a loss of enzyme functioning. (4)

Dendrites Cytoplasmic extensions of the cell body of a neuron. They carry impulses from the area of stimulation to the cell body. (23)

Denitrification The conversion by denitrifying bacteria of nitrites and nitrates into nitrogen gas. (41)

Denitrifying Bacteria Bacteria which take soil nitrogen, usable to plants, and convert it to unusable nitrogen gas. (41)

Density-Dependent Factors Factors that control population growth which are influenced by population size. (43)

Density-Independent Factors Factors that control population growth which are not affected by population size. (43)

Deoxyribonucleic Acid (DNA) Double-stranded polynucleotide comprised of deoxyribose (a sugar), phosphate, and four bases (adenine, guanine, cytosine, and thymine). Encoded in the sequence of nucleotides are the instructions for making proteins. DNA is the genetic material in all organisms except certain viruses. (14)

Depolarization A decrease in the potential difference (voltage) across the plasma membrane of a cell typically due to an increase in the movement of sodium ions into the cell. Acts to excite a target cell. (23)

Dermal Bone Bones of vertebrates that form within the dermal layer of the skin, such as the scales of fishes and certain bones of the skull. (26)

Dermal Tissue System In plants, the epidermis in primary growth, or the periderm in secondary growth. (18)

Dermis In animals, layer of cells below the epidermis in which connective tissue predominates. Embedded within it are vessels, various glands, smooth muscle, nerves, and follicles. (26)

Desert Biome characterized by intense solar radiation, very little rainfall, and high winds. (40)

Detrivore Organism that feeds on detritus, dead organisms or their parts, and living organisms' waste. (41)

Deuterostome One path of development exhibited by coelomate animals (e.g., echinoderms and chordates). (39)

Diabetes Mellitus A disease caused by a deficiency of insulin or its receptor, preventing glucose from being absorbed by the cells. (25)

Diaphragm A sheet of muscle that separates the thoracic cavity from the abdominal wall. (29)

Diastolic Pressure The second number of a blood pressure reading; the lowest pressure in the arteries just prior to the next heart contraction. (28)

Diatoms are golden-brown algae that are distinguished most dramatically by their intricate silica shells. (37)

Dicotyledonae (Dicots) One of the two classes of flowering plants, characterized by having seeds with two cotyledons, flower parts in 4s or 5s, net-veined leaves, one main root, and vascular bundles in a circular array within the stem. (Compare with Monocotylenodonae). (18)

Diffusion Tendency of molecules to move from a region of higher concentration to a region of lower concentration, until they are uniformly dispersed. (7)

Digestion The process by which food particles are disassembled into molecules small enough to be absorbed into the organism's cells and tissues. (27)

Digestive System System of specialized organs that ingests food, converts nutrients to a form that can be distributed throughout the animal's body, and eliminates undigested residues. (27)

Digestive Tract Tubelike channel through which food matter passes from its point of ingestion at the mouth to the elimination of indigestible residues from the anus. (27)

Dihybrid Cross A mating between two individuals that differ in two genetically-determined traits. (12)

Dimorphism Presence of two forms of a trait within a population, resulting from diversifying selection. (33)

Dinoflagellates Single-celled photosynthesizers that have two flagella. They are members of the pyrophyta, phosphorescent algae that sometimes cause red tide, often synthesizing a neurotoxin that accumulates in plankton eaters, causing paralytic shellfish poisoning in people who eat the shellfish. (37)

Dioecious Plants that produce either male or female reproductive structures but never both. (38)

Diploid Having two sets of homologous chromosomes. Often written 2N. (10, 13)

Directional Selection The steady shift of phenotypes toward one extreme. (33)

Discontinuous Variation An inheritance pattern in which the phenomenon of all possible phenotypes fall into distinct categories. (Compare with continuous variation). (12)

Displays The signals that form the language by which animals communicate. These signals are species specific and stereotyped and may be visual, auditory, chemical, or tactile. (44)

Disruptive Coloration Coloration that disguises the shape of an organism by breaking up its outline. (42)

Disruptive Selection The steady shift toward more than one extreme phenotype due to the elimination of intermediate phenotypes as has occurred among African swallowtail butterflies whose members resemble more than one species of distasteful butterfly. (33)

Divergent Evolution The emergence of new species as branches from a single ancestral lineage. (33)

Diversifying Selection The increasing frequency of extreme phenotypes because individuals with average phenotypes die off. (33)

Diving Reflex Physiological response that alters the flow of blood in the body of diving mammals that allows the animal to maintain high levels of activity without having to breathe. (29)

Division (or Phylum) A level of the taxonomic hierarchy that groups together members or related classes. (1)

DNA (see **Deoxyribonucleic Acid**)

DNA Cloning The amplification of a particular DNA by use of a growing population of bacteria. The DNA is initially taken up by a bacterial cell—usually as a plasmid—and then replicated along with the bacteria's own DNA. (16)

DNA Fingerprint The pattern of DNA fragments produced after treating a sample of DNA with a particular restriction enzyme and separating the fragments by gel electrophoresis. Since different members of a population have DNA with a different nucleotide sequence, the pattern of DNA fragments produced by this method can be used to identify a particular individual. (16)

DNA Ligase The enzyme that covalently joins DNA fragments into a continuous DNA strand. The enzyme is used in a cell during replication to seal newly-synthesized fragments and by biotechnologists to form recombinant DNA molecules from separate fragments. (14, 16)

DNA Polymerase Enzyme responsible for replication of DNA. It assembles free nucleotides, aligning them with the complementary ones in the unpaired region of a single strand of DNA template. (14)

Dominant The form of an allele that masks the presence of other alleles for the same trait. (12)

Dormancy A resting period, such as seed dormancy in plants or hibernation in animals, in which organisms maintain reduced metabolic rates. (21)

Dorsal In anatomy, the back of an animal. (39)

Double Blind Test A clinical trial of a drug in which neither the human subjects or the researchers know who is receiving the drug or placebo. (2)

Down Syndrome Genetic disorder in humans characterized by distinct facial appearance and mental retardation, resulting from an extra copy of chromosome number 21 (trisomy 21) in each cell. (11, 17)

Duodenum First part of the human small intestine in which most digestion of food occurs. (27)

Duplication The repetition of a segment of a chromosome. (13)

◄ E ►

Ecdysis Molting process by which an arthropod periodically discards its exoskeleton and replaces it with a larger version. The process is controlled by the hormone ecydysone. (39)

Ecdysone An insect steroid hormone that triggers molting and metamorphosis. (15)

Echinodermata A phylum composed of animals having an internal skeleton made of many small calcium carbonate plates which have jutting spines. Includes sea stars, sea urchins, etc. (39)

Echolocation The use of reflected sound waves to help guide an animal through its environment and/or locate objects. (24)

Ecological Equivalent Organisms that occupy similar ecological niches in different regions or ecosystems of the world. (41)

Ecological Niche The habitat, functional role(s), requirements for environmental resources and tolerance ranges for each abiotic condition in relation to an organism. (41)

Ecological Pyramid Illustration showing the energy content, numbers of organisms, or biomass at each trophic level. (41)

Ecology The branch of biology that studies interactions among organisms as well as the interactions of organisms and their physical environment. (40)

Ecosystem Unit comprised of organisms interacting among themselves and with their physical environment. (41)

Ecotypes Populations of a single species with different, genetically fixed tolerance ranges. (41)

Ectoderm In animals, the outer germ cell layer of the gastrula. It gives rise to the nervous system and integument. (32)

Ectotherms Animals that lack an internal mechanism for regulating body temperature. "Cold-blooded" animals. (28)

Edema Swelling of a tissue as the result of an accumulation of fluid that has moved out of the blood vessels. (28)

Effectors Muscle fibers and glands that are activated by neural stimulation. (23)

Efferent (Motor) Nerves The nerves that carry messages from the central nervous system to the effectors, the muscles, and glands. They are divided into two systems: somatic and autonomic. (23)

Egg Female gamete, also called an ovum. A fertilized egg is the product of the union of female and male gametes (egg and sperm cells). (32)

Electrocardiogram (EKG) Recording of the electrical activity of the heart, which is used to diagnose various types of heart problems. (28)

Electron Acceptor Substances that are capable of accepting electrons transferred from an electron donor. For example, molecular oxygen (O_2) is the terminal electron acceptor during respiration. Electron acceptors also receive electrons from chlorophyll during photosynthesis. Electron acceptors may act as part of an electron transport system by transferring the electrons they receive to another substance. (8, 9)

Electron Carrier Substances (such as NAD^+ and FAD) that transport electrons from one step of a metabolic pathway to the next or from metabolic reactions to biosynthetic reactions. (8, 9)

Electrons Negatively charged particles that orbit the atomic nucleus. (3)

Electron Transport System Highly organized assembly of cytochromes and other proteins which transfer electrons. During transport, which occurs within the inner membranes of mitochondria and chloroplasts, the energy extracted from the electrons is used to make ATP. (8, 9)

Electrophoresis A technique for separating different molecules on the basis of their size and/or electric charge. There are various ways the technique is used. In gel electrophoresis, proteins or DNA fragments are driven through a porous gel by their charge, but become separated according to size; the larger the molecule, the slower it can work its way through the pores in the gel, and the less distance it travels along the gel. (16)

Element Substance composed of only one type of atom. (3)

Embryo An organism in the early stages of development, beginning with the first division of the zygote. (32)

Embryo Sac The fully developed female gametophyte within the ovule of the flower. (20)

Emigration Individuals permanently leaving an area or population. (43)

Endergonic Reactions Chemical reactions that require energy input from another source in order to occur. (6)

Endocrine Glands Ductless glands, which secrete hormones directly into surrounding tissue fluids and blood vessels for distribution to the rest of the body by the circulatory system. (25)

Endocytosis A type of active transport that imports particles or small cells into a cell. There are two types of endocytic processes: phagocytosis, where large particles are ingested by the cell, and pinocytosis, where small droplets are taken in. (7)

Endoderm In animals, the inner germ cell layer of the gastrula. It gives rise to the digestive tract and associated organs and to the lungs. (32)

Endodermis The innermost cylindrical layer of cortex surrounding the vascular tissues of the root. The closely pressed cells of the endodermis have a waxy band, forming a waterproof layer, the Casparian strip. (18)

Endogenous Plant responses that are controlled internally, such as biological clocks controlling flower opening. (21)

Endometrium The inner epithelial layer of the uterus that changes markedly with the uterine (menstrual) cycle in preparation for implantation of an embryo. (31)

Endoplasmic Reticulum (ER) An elaborate system of folded, stacked and tubular membranes contained in the cytoplasm of eukaryotic cells. (5)

Endorphins (Endogenous Morphinelike Substances) A class of peptides released from nerve cells of the limbic system of the brain that can block perceptions of pain and produce a feeling of euphoria. (23)

Endoskeleton The internal support structure found in all vertebrates and a few invertebrates (sponges and sea stars). (26)

Endosperm Nutritive tissue in plant embryos and seeds. (20)

Endosperm Mother Cell A binucleate cell in the embryo sac of the female gametophyte, occurring in the ovule of the ovary in angiosperms. Each nucleus is haploid; after fertilization, nutritive endosperm develops. (20)

Endosymbiosis Theory A theory to explain the development of complex eukaryotic cells by proposing that some organelles once were free-living prokaryotic cells that then moved into another larger such cell, forming a beneficial union with it. (5)

Endotherms Animals that utilize metabolically produced heat to maintain a constant, elevated body temperature. "Warm-blooded" animals. (28)

End Product The last product in a metabolic pathway. Typically a substance, such as an amino acid or a nucleotide, that will be used as a monomer in the formation of macromolecules. (6)

Energy The ability to do work. (6)

Entropy Energy that is not available for doing work; measure of disorganization or randomness. (6)

Environmental Resistance The factors that eventually limit the size of a population. (43)

Enzyme Biological catalyst; a protein molecule that accelerates the rate of a chemical reaction. (6)

Eosiniphil A type of phagocytic white blood cell. (28)

Epicotyl The portion of the embryo of a dicot plant above the cotyledons. The epicotyl gives rise to the shoot. (20)

Epidermis In vertebrates, the outer layer of the skin, containing superficial layers of dead cells produced by the underlying living epithelial cells. In plants, the outer layer of cells covering leaves, primary stem, and primary root. (26, 18)

Epididymis Mass of convoluted tubules attached to each testis in mammals. After leaving the testis, sperm enter the tubules where they finish maturing and acquire motility. (31)

Epiglottis A flap of tissue that covers the glottis during swallowing to prevent food and liquids from entering the lower respiratory tract. (29)

Epinephrine (Adrenalin) Substance that serves both as an excitatory neurotransmitter released by certain neurons of the CNS and as a hormone released by the adrenal medulla that increases the body's ability to combat a stressful situation. (25)

Epipelagic Zone The lighted upper ocean zone, where photosynthesis occurs; large populations of phytoplankton occur in this zone. (40)

Epiphyseal Plates The action centers for ossification (bone formation). (26)

Epistasis A type of gene interaction in which a particular gene blocks the expression of another gene at another locus. (12)

Epithelial Tissue Continuous sheets of tightly packed cells that cover the body and line its tracts and chambers. Epithelium is a fundamental tissue type in animals. (22)

Erythrocytes Red blood cells. (28)

Erythropoietin A hormone secreted by the kidney which stimulates the formation of erythrocytes by the bone marrow. (28)

Essential Amino Acids Eight amino acids that must be acquired from dietary protein. If even one is missing from the human diet, the synthesis of proteins is prevented. (27)

Essential Fatty Acids Linolenic and linoleic acids, which are required for phospholipid construction and must be acquired from a dietary source. (27)

Essential Nutrients The 16 minerals essential for plant growth, divided into two groups: macronutrients, which are required in large quantities, and micronutrients, which are needed in small amounts. (19)

Estrogen A female sex hormone secreted by the ovaries when stimulated by pituitary gonadotrophins. (31)

Estuaries Areas found where rivers and streams empty into oceans, mixing fresh water with salt water. (40)

Ethology The study of animal behavior. (44)

Ethylene Gas A plant hormone that stimulates fruit ripening. (21)

Etiolation The condition of rapid shoot elongation, small underdeveloped leaves, bent shoot-hook, and lack of chlorophyll, all due to lack of light. (21)

Eubacteria Typical procaryotic bacteria with peptidoglycan in their cell walls. The majority of monerans are eubacteria. (36)

Eukaryotic Referring to organisms whose cellular anatomy includes a true nucleus with a nuclear envelope, as well as other membrane-bound organelles. (5)

Eusocial Species Social species that have sterile workers, cooperative care of the young, and an overlap of generations so that the colony labor is a family affair. (44)

Eutrophication The natural aging process of lakes and ponds, whereby they become marshes and, eventually, terrestrial environments.

Evolution A process whereby the characteristics of a species change over time, eventually leading to the formation of new species that go about life in new ways. (33)

Evolutionarily Stable Strategy (ESS) A behavioral strategy or course of action that depends on what other members of the population are doing. By definition, an ESS cannot be replaced by any other strategy when most of the members of the population have adopted it. (44)

Excitatory Neurons Neurons that stimulate their target cells into activity. (23)

Excretion Removal of metabolic wastes from an organism. (28)

Excretory System The organ system that eliminates metabolic wastes from the body. (28)

Exergonic Reactions Chemical reactions that occur spontaneously with the release of energy. (6)

Exocrine Glands Glands which secrete their products through ducts directly to their sites of action, e.g., tear glands. (26)

Exocytosis A form of active transport used by cells to move molecules, particles, or other cells contained in vesicles across the plasma membrane to the cell's environment. (5)

Exogenous Plant responses that are controlled externally, or by environmental conditions. (21)

Exons Structural gene segments that are transcribed and whose genetic information is subsequently translated into protein. (15)

Exoskeletons Hard external coverings found in some animals (e.g., lobsters, insects) for protection, support, or both. Such organisms grow by the process of molting. (26)

Exploitative Competition A competition in which one species manages to get more of a resource, thereby reducing supplies for a competitor. (42)

Exponential Growth An increase by a fixed percentage in a given time period; such as population growth per year. (43)

Extensor Muscle A muscle which, when contracted, causes a part of the body to straighten at a joint. (26)

External Fertilization Fertilization of an egg outside the body of the female parent. (31)

Extinction The loss of a species. (33)

Extracellular Digestion Digestion occurring outside the cell; occurs in bacteria, fungi, and multicellular animals. (27)

Extracellular Matrix Layer of extracellular material residing just outside a cell. (5)

◀ F ▶

F_1 First filial generation. The first generation of offspring in a genetic cross. (12)

F_2 Second filial generation. The offspring of an F_1 cross. (12)

Facilitated Diffusion The transport of molecules into cells with the aid of "carrier" proteins embedded in the plasma membrane. This carrier-assisted transport does not require the expenditure of energy by the cell. (7)

FAD Flavin adenine dinucleotide. A coenzyme that functions as an electron carrier in metabolic reactions. When it is reduced to $FADH_2$, this molecule becomes a cellular energy source. (9)

Family A level of the taxonomic hierarchy that groups together members of related genera. (1)

Fast-Twitch Fibers Skeletal muscle fibers that depend on anaerobic metabolism to produce ATP rapidly, but only for short periods of time before the onset of fatigue. Fast-twitch fibers generate greater forces for shorter periods than slow-twitch fibers. (9)

Fat A triglyceride consisting of three fatty acids joined to a glycerol. (4)

Fatty Acid A long unbranched hydrocarbon chain with a carboxyl group at one end. Fatty acids lacking a double bond are said to be saturated. (4)

Fauna The animals in a particular region.

Feedback Inhibition (Negative Feedback) A mechanism for regulating enzyme activity by temporarily inactivating a key enzyme in a biosynthetic pathway when the concentration of the end product is elevated. (6)

Fermentation The direct donation of the electrons of NADH to an organic compound without their passing through an electron transport system. (9)

Fertility Rate In humans, the average number of children born to each woman between 15 and 44 years of age. (43)

Fertilization The process in which two haploid nuclei fuse to form a zygote. (32)

Fetus The term used for the human embryo during the last seven months in the uterus. During the fetal stage, organ refinement accompanies overall growth. (32)

Fibrinogen A rod-shaped plasma protein that, converted to fibrin, generates a tangled net of fibers that binds a wound and stops blood loss until new cells replace the damaged tissue. (28)

Fibroblasts Cells found in connective tissues that secrete the extracellular materials of the connective tissue matrix. These cells are easily isolated from connective tissues and are widely used in cell culture. (22)

Fibrous Root System Many approximately equal-sized roots; monocots are characterized by a fibrous root system. Also called diffuse root system. (18)

Filament The stalk of a stamen of angiosperms, with the anther at its tip. Also, the threadlike chain of cells in some algae and fungi. (20)

Filamentous Fungus Multicellular members of the fungus kingdom comprised mostly of living threads (hyphae) that grow by division of cells at their tips (see molds). (37)

Filter Feeders Aquatic animals that feed by straining small food particles from the surrounding water. (27, 39)

Fitness The relative degree to which an individual in a population is likely to survive to reproductive age and to reproduce. (33)

Fixed Action Patterns Motor responses that may be triggered by some environmental stimulus, but once started can continue to completion without external stimuli. (44)

Flagella Cellular extensions that are longer than cilia but fewer in number. Their undulations propel cells like sperm and many protozoans, through their aqueous environment. (5)

Flexor Muscle A muscle which, when contracted, causes a part of the body to bend at a joint. (26)

Flora The plants in a particular region. (21)

Florigen Proposed A chemical hormone that is produced in the leaves and stimulates flowering. (21)

Fluid Mosaic Model The model proposes that the phospholipid bilayer has a viscosity similar to that of light household oil and that globular proteins float like icebergs within this bilayer. The now favored explanation for the architecture of the plasma membrane. (5)

Follicle (Ovarian) A chamber of cells housing the developing oocytes. (31)

Food Chain Transfers of food energy from organism to organism, in a linear fashion. (41)

Food Web The map of all interconnections between food chains for an ecosystem. (41)

Forest Biomes Broad geographic regions, each with characteristic tree vegetation: 1) tropical rain forests (lush forests in a broad band around the equator), 2) deciduous forests (trees and shrubs drop their leaves during unfavorable seasons), 3) coniferous forest (evergreen conifers). (40)

Fossil Record An entire collection of remains from which paleontologists attempt to reconstruct the phylogeny, anatomy, and ecology of the preserved organisms. (34)

Fossils The preserved remains of organisms from a former geologic age. (34)

Fossorial Living underground.

Founder Effect The potentially dramatic difference in allele frequency of a small founding population as compared to the original population. (33)

Founder Population The individuals, usually few, that colonize a new habitat. (33)

Frameshift Mutation The insertion or deletion of nucleotides in a gene that throws off the reading frame. (14)

Free Radical Atom or molecule containing an unpaired electron, which makes it highly reactive. (3)

Freeze-Fracture Technique in which cells are frozen into a block which is then struck with a knife blade that fractures the block in two. Fracture planes tend to expose the center of membranes for EM examination. (5)

Fronds The large leaf-like structures of ferns. Unlike true leaves, fronds have an apical meristem and clusters of sporangia called sori. (38)

Fruit A mature plant ovary (flower) containing seeds with plant embryos. Fruits protect seeds and aid in their dispersal. (20)

Fruiting Body A spore-producing structure that extends upward in an elevated position from the main mass of a mold or slime mold. (37)

FSH Follicle stimulating hormone. A hormone secreted by the anterior pituitary that prepares a female for ovulation by stimulating the primary follicle to ripen or stimulates spermatogenesis in males. (31)

Functional Groups Accessory chemical entities (e.g., —OH, $—NH_2$, $—CH_3$), which help determine the identity and chemical properties of a compound. (4)

Fundamental Niche The potential ecological niche of a species, including all factors affecting that species. The fundamental niche is usually never fully utilized. (41)

Fungus Yeast, mold, or large filamentous mass forming macroscopic fruiting bodies, such as mushrooms. All fungi are eukaryotic nonphotosynthetic heterotrophics with cell walls. (37)

◄ G ►

G_1 Stage The first of three consecutive stages of interphase. During G_1, cell growth and normal functions occur. The duration of this stage is most variable. (10)

G_2 Stage The final stage of interphase in which the final preparations for mitosis occur. (10)

Gallbladder A small saclike structure that stores bile salts produced by the liver. (27)

Gamete A haploid reproductive cell—either a sperm or an egg. (10)

Gas Exchange Surface Surface through which gases must pass in order to enter or leave the body of an animal. It may be the plasma membrane of a protistan or the complex tissues of the gills or the lungs in multicellular animals. (29)

Gastrovascular Cavity In cnidarians and flatworms, the branched cavity with only one opening. It functions in both digestion and transport of nutrients. (39)

Gastrula The embryonic stage formed by the inward migration of cells in the blastula. (32)

Gastrulation The process by which the blastula is converted into a gastrula having three germ layers (ectoderm, mesoderm, and endoderm). (32)

Gated Ion Channels Most passageways through a plasma membrane that allow ions to pass contain "gates" that can occur in either an open or a closed conformation. (7, 23)

Gel Electrophoresis (See **Electrophoresis**)

Gene Pool All the genes in all the individuals of a population. (33)

Gene Regulatory Proteins Proteins that bind to specific sites in the DNA and control the transcription of nearby genes. (15)

Genes Discrete units of inheritance which determine hereditary traits. (12, 14)

Gene Therapy Treatment of a disease by alteration of the person's genotype, or the genotype of particular affected cells. (17)

Genetic Carrier A heterozygous individual who shows no evidence of a genetic disorder but, because they possess a recessive allele for a disorder, can pass the mutant gene on to their offspring. (17)

Genetic Code The correspondence between the various mRNA triplets (codons, e.g., UGC) and the amino acid that the triplet specifies (e.g., cysteine). The genetic code includes 64 possible three-letter words that constitute the genetic language for protein synthesis. (14)

Genetic Drift Random changes in allele frequency that occur by chance alone. Occurs primarily in small populations. (33)

Genetic Engineering The modification of a cell or organism's genetic composition according to human design. (16)

Genetic Equilibrium A state in which allele frequencies in a population remain constant from generation to generation. (33)

Genetic Mapping Determining the locations of specific genes or genetic markers along particular chromosomes. This is typically accomplished using crossover frequencies; the more often alleles of two genes are separated during crossing over, the greater the distance separating the genes. (13)

Genetic Recombination The reshuffling of genes on a chromosome caused by breakage of DNA and its reunion with the DNA of a homologoue. (11)

Genome The information stored in all the DNA of a single set of chromosomes. (17)

Genotype An individual's genetic makeup. (12)

Genus Taxonomic group containing related species. (1)

Geologic Time Scale The division of the earth's 4.5 billion-year history into eras, periods, and epochs based on memorable geologic and biological events. (35)

Germ Cells Cells that are in the process of or have the potential to undergo meiosis and form gametes. (11, 31)

Germination The sprouting of a seed, beginning with the radicle of the embryo breaking through the seed coat. (21)

Germ Layers Collective name for the endoderm, ectoderm, and mesoderm, from which all the structures of the mature animal develop. (32)

Gibberellins More than 50 compounds that promote growth by stimulating both cell elongation and cell division. (21)

Gills Respiratory organs of aquatic animals. (29)

Globin The type of polypeptide chains that make up a hemoglobin molecule.

Glomerular Filtration The process by which fluid is filtered out of the capillaries of the glomerulus into the proximal end of the nephron. Proteins and blood cells remain behind in the bloodstream. (28)

Glomerulus A capillary bundle embedded in the double-membraned Bowman's capsule, through which blood for the kidney first passes. (28)

Glottis Opening leading to the larynx and lower respiratory tract. (29)

Glucagon A hormone secreted by the Islets of Langerhans that promotes glycogen breakdown to glucose. (25)

Glucocorticoids Steroid hormones which regulate sugar and protein metabolism. They are secreted by the adrenal cortex. (25)

Glycogen A highly branched polysaccharide consisting of glucose monomers that serves as a storage of chemical energy in animals. (4)

Glycolysis Cleavage, releasing energy, of the six-carbon glucose molecule into two molecules of pyruvic acid, each containing three carbons. (9)

Glycoproteins Proteins with covalently-attached chains of sugars. (5)

Glycosidic Bond The covalent bond between individual molecules in carbohydrates. (4)

Golgi Complex A system of flattened membranous sacs, which package substances for secretion from the cell. (5)

Gonadotropin-Releasing Hormone (GnRH) Hypothalmic hormone that controls the secretion of the gonadotropins FSH and LH. (31)

Gonadotropins Two anterior pituitary hormones which act on the gonads. Both FSH (follicle-stimulating hormone) and LH (luteinizing hormone) promote gamete development and stimulate the gonads to produce sex hormones. (25)

Gonads Gamete-producing structures in animals: ovaries in females, testes in males. (31)

Grasslands Areas of densely packed grasses and herbaceous plants. (40)

Gravitropisms (Geotropisms) Changes in plant growth caused by gravity. Growth away from gravitational force is called negative gravitropism; growth toward it is positive. (21)

Gray Matter Gray-colored neural tissue in the cerebral cortex of the brain and in the butterfly-shaped interior of the spinal cord. Composed of nonmyelinated cell bodies and dendrites of neurons. (23)

Greenhouse Effect The trapping of heat in the Earth's troposphere, caused by increased levels of carbon dioxide near the Earth's surface; the carbon dioxide is believed to act like glass in a greenhouse, allowing light to reach the Earth, but not allowing heat to escape. (41)

Ground Tissue System All plant tissues except those in the dermal and vascular tissues. (18)

Growth An increase in size, resulting from cell division and/or an increase in the volume of individual cells. (10)

Growth Hormone (GH) Hormone produced by the anterior pituitary; stimulates protein synthesis and bone elongation. (25)

Growth Ring In plants with secondary growth, a ring formed by tracheids and/or vessels with small lumens (late wood) during periods of unfavorable conditions; apparent in cross section. (18)

Guard Cells Specialized epidermal plant cells that flank each stomated pore of a leaf. They regulate the rate of gas diffusion and transpiration. (18)

Guild Group of species with similar ecological niches. (41)

Guttation The forcing of water and mineral completely out to the tips of leaves as a result of positive root pressure. (19)

Gymnosperms The earliest seed plants, bearing naked seeds. Includes the pines, hemlocks, and firs. (38)

◀ H ▶

Habitat The place or region where an organism lives. (41)

Habituation The phenomenon in which an animal ceases to respond to a repetitive stimulus. (23, 44)

Hair Cells Sensory receptors of the inner ear that respond to sound vibration and bodily movement. (24)

Half-Life The time required for half the mass of a radioactive element to decay into its stable, non-radioactive form. (3)

Haplodiploidy A genetic pattern of sex determination in which fertilized eggs develop into females and non-fertilized eggs develop into males (as occurs among bees and wasps). (44)

Haploid Having one set of chromosomes per cell. Often written as 1N. (10)

Hardy-Weinberg Law The maintenance of constant allele frequencies in a population from one generation to the next when certain conditions are met. These conditions are the absence of mutation and migration, random mating, a large population, and an equal chance of survival for all individuals. (33)

Haversian Canals A system of microscopic canals in compact bone that transport nutrients to and remove wastes from osteocytes. (26)

Heart An organ that pumps blood (or hemolymph in arthropods) through the vessels of the circulatory system. (28)

Helper T Cells A class of T cells that regulate immune responses by recognizing and activating B cells and other T cells. (30)

Hemocoel In arthropods, the unlined spaces into which fluid (hemolymph) flows when it leaves the blood vessels and bathes the internal organs. (28)

Hemoglobin The iron-containing blood protein that temporarily binds O_2 and releases it into the tissues. (4, 29)

Hemophilia A genetic disorder determined by a gene on the X chromosome (an X-linked trait) that results from the failure of the blood to form clots. (13)

Herbaceous Plants having only primary growth and thus composed entirely of primary tissue. (18)

Herbivore An organism, usually an animal, that eats primary producers (plants). (42)

Herbivory The term for the relationship of a secondary consumer, usually an animal, eating primary producers (plants). (42)

Heredity The passage of genetic traits to offspring which consequently are similar or identical to the parent(s). (12)

Hermaphrodites Animals that possess gonads of both the male and the female. (31)

Heterosporous Higher vascular plants producing two types of spores, a megaspore which grows into a female gametophyte and a microspore which grows into a male gametophyte. (38)

Heterozygous A term applied to organisms that possess two different alleles for a trait. Often, one allele (A) is dominant, masking the presence of the other (a), the recessive. (12)

High Intertidal Zone In the intertidal zone, the region from mean high tide to around just below sea level. Organisms are submerged about 10% of the time. (40)

Histones Small basic proteins that are complexed with DNA to form nucleosomes, the basic structural components of the chromatin fiber. (14)

Homeobox That part of the DNA sequence of homeotic genes that is similar (homologous) among diverse animal species. (32)

Homeostasis Maintenance of fairly constant internal conditions (e.g., blood glucose level, pH, body temperature, etc.) (22)

Homeotic Genes Genes whose products act during embryonic development to affect the spatial arrangement of the body parts. (32)

Hominids Humans and the various groups of extinct, erect-walking primates that were either our direct ancestors or their relatives. Includes the various species of *Homo* and *Australopithecus*. (34)

Homo the genus that contains modern and extinct species of humans. (34)

Homologous Structures Anatomical structures that may have different functions but develop from the same embryonic tissues, suggesting a common evolutionary origin. (34)

Homologues Members of a chromosome pair, which have a similar shape and the same sequence of genes along their length. (10)

Homoplasy (see **Analogous Structures**)

Homosporous Plants that manufacture only one type of spore, which develops into a gametophyte containing both male and female reproductive structures. (38)

Homozygous A term applied to an organism that has two identical alles for a particular trait. (12)

Hormones Chemical messengers secreted by ductless glands into the blood that direct tissues to change their activities and correct imbalances in body chemistry. (25)

Host The organism that a parasite lives on and uses for food. (42)

Human Chorionic Gonadotropin (HCG) A hormone that prevents the corpus luteum from degenerating, thereby maintaining an adequate level of progesterone during pregnancy. It is produced by cells of the early embryo. (25)

Human Immunodeficiency Virus (HIV) The infectious agent that causes AIDS, a disease in which the immune system is seriously disabled. (30, 36)

Hybrid An individual whose parents possess different genetic traits in a breeding experiment or are members of different species. (12)

Hybridization Occurs when two distinct species mate and produce hybrid offspring. (33)

Hybridoma A cell formed by the fusion of a malignant cell (a myeloma) and an antibody-producing lymphocyte. These cells proliferate indefinitely and produce monoclonal antibodies. (30)

Hydrogen Bonds Relatively weak chemical bonds formed when two molecules share an atom of hydrogen. (3)

Hydrologic Cycle The cycling of water, in various forms, through the environment, from Earth to atmosphere and back to Earth again. (41)

Hydrolysis Splitting of a covalent bond by donating the H^+ or OH^- of a water molecule to the two components. (4)

Hydrophilic Molecules Polar molecules that are attracted to water molecules and readily dissolve in water. (3)

Hydrophobic Interaction When nonpolar molecules are "forced" together in the presence of a polar solvent, such as water. (3)

Hydrophobic Molecules Nonpolar substances, insoluble in water, which form aggregates to minimize exposure to their polar surroundings. (3)

Hydroponics The science of growing plants in liquid nutrient solutions, without a solid medium such as soil. (19)

Hydrosphere That portion of the Earth composed of water. (40)

Hydrostatic Skeletons Body support systems found usually in underwater animals (e.g., marine worms). Body shape is protected against gravity and other physical forces by internal hydrostatic pressure produced by contracting muscles encircling their closed, fluid-filled chambers. (26)

Hydrothermal Vents Fissures in the ocean floor where sea water becomes superheated. Chemosynthetic bacteria that live in these vents serve as the autotrophs that support a diverse community of ocean-dwelling organisms. (8)

Hyperpolarization An increase in the potential difference (voltage) across the plasma membrane of a cell typically due to an increase in the movement of potassium ions out of the cell. Acts to inhibit a target cell. (23)

Hypertension High blood pressure (above about 130/90). (28)

Hypertonic Solutions Solutions with higher solute concentrations than found inside the cell. These cause a cell to lose water and shrink. (7)

Hypervolume In ecology, a multidimensional area which includes all factors in an organism's ecological niche, or its' potential niche. (41)

Hypocotyl Portion of the plant embryo below the cotyledons. The hypocotyl gives rise to the root and, very often, to the lower part of the stem. (20)

Hypothalamus The area of the brain below the thalamus that regulates body temperature, blood pressure, etc. (25)

Hypothesis A tentative explanation for an observation or a phenomenon, phrased so that it can be tested by experimentation. (2)

Hypotonic Solutions Solutions with lower solute concentrations than found inside the cell. These cause a cell to accumulate water and swell. (7)

◀ I ▶

Immigration Individuals permanently moving into a new area or population. (43)

Immune System A system in vertebrates for the surveillance and destruction of disease-causing microorganisms and cancer cells. Composed of lymphocytes, particularly B cells and T cells, and triggered by the introduction of antigens into the body which makes the body, upon their destruction, resistant to a recurrence of the same disease. (30)

Immunoglobulins (IGs) Antibody molecules. (30)

Imperfect Flowers Flowers that contain either stamens or carpels, making them male or female flowers, respectively. (20)

Imprinting A type of learning in which an animal develops an association with an object after exposure to the object during a critical period early in its life. (44)

Inbreeding When individuals mate with close relatives, such as brothers and sisters. May occur when population sizes drastically shrink and results in a decrease in genetic diversity. (33)

Incomplete Digestive Tract A digestive tract with only one opening through which food is taken in and residues are expelled. (27)

Incomplete (Partial) Dominance A phenomenon in which heterozygous individuals are phenotypically distinguishable from either homozygous type. (12)

Incomplete Flower Flowers lacking one or more whorls of sepals, petals, stamen, or pistils. (20)

Independent Assortment The shuffling of members of homologous chromosome pairs in meiosis I. As a result, there are new chromosome combinations in the daughter cells, which later produce offspring with random mixtures of traits from both parents. (11, 12)

Indoleatic Acid (IAA) An auxin responsible for many plant growth responses including apical dominance, a growth pattern in which shoot tips prevent axillary buds from sprouting. (21)

Induction The process in which one embryonic tissue induces another tissue to differentiate along a pathway that it would not otherwise have taken. (32) Stimulation of transcription of a gene in an operon. Occurs when the repressor protein is unable to bind to the operator. (15)

Inflammation A body strategy initiated by the release of chemicals following injury or infection which brings additional blood with its protective cells to the injured area. (30)

Inhibitory Neurons Neurons that oppose a response in the target cells. (23)

Inhibitory Neurotransmitters Substances released from inhibitory neurons where they synapse with the target cell. (23)

Innate Behavior Actions that are under fairly precise genetic control, typically species-specific, highly stereotyped, and that occur in a complete form the first time the stimulus is encountered. (44)

Insight Learning The sudden solution to a problem without obvious trial-and-error procedures. (44)

Insulin One of the two hormones secreted by endocrine centers called Islets of Langerhans; promotes glucose absorption, utilization, and storage. Insulin is secreted by them when the concentration of glucose in the blood begins to exceed the normal level. (25)

Integumentary System The body's protective external covering, consisting of skin and subcutaneous tissue. (26)

Integuments Protective covering of the ovule. (20)

Intercellular Junctions Specialized regions of cell-cell contact between animal cells. (5)

Intercostal Muscles Muscles that lie between the ribs in humans whose contraction expands the thoracic cavity during breathing. (29)

Interference Competition One species' direct interference by another species for the same limited resource; such as aggressive animal behavior. (42)

Internal Fertilization Fertilization of an egg within the body of the female. (31)

Interneurons Neurons situated entirely within the central nervous system. (23)

Internodes The portion of a stem between two nodes. (18)

Interphase Usually the longest stage of the cell cycle during which the cell grows, carries out normal metabolic functions, and replicates its DNA in preparation for cell division. (10)

Interstitial Cells Cells in the testes that produce testosterone, the major male sex hormone. (31)

Interstitial Fluid The fluid between and surrounding the cells of an animal; the extracellular fluid. (28)

Intertidal Zone The region of beach exposed to air between low and high tides. (40)

Intracellular Digestion Digestion occurring inside cells within food vacuoles. The mode of digestion found in protists and some filter-feeding animals (such as sponges and clams). (27)

Intraspecific Competition Individual organisms of one species competing for the same limited resources in the same habitat, or with overlapping niches. (42)

Intrinsic Rate of Increase (r_m) the maximum growth rate of a population under conditions of maximum birth rate and minimum death rate. (43)

Introns Intervening sequences of DNA in the middle of structural genes, separating exons. (15)

Invertebrates Animals that lack a vertebral column, or backbone. (39)

Ion An electrically charged atom created by the gain or loss of electrons. (3)

Ionic Bond The noncovalent linkage formed by the attraction of oppositely charged groups. (3)

Islets of Langerhans Clusters of endocrine cells in the pancreas that produce insulin and glucagon. (25)

Isolating Mechanisms Barriers that prevent gene flow between populations or among segments of a single population. (33)

Isotopes Atoms of the same element having a different number of neutrons in their nucleus. (3)

Isotonic Solutions Solutions in which the solute concentration outside the cell is the same as that inside the cell. (7)

◄ J ►

Joints Structures where two pieces of a skeleton are joined. Joints may be flexible, such as the knee joint of the human leg or the joints between segments of the exoskeleton of the leg of an insect, or inflexible, such as the joints (sutures) between the bones of the skull. (26)

J-Shaped Curve A curve resulting from exponential growth of a population. (43)

◄ K ►

Karyotype A visual display of an individual's chromosomes. (10)

Kidneys Paired excretory organs which, in humans, are fist-sized and attached to the lower spine. In vertebrates, the kidneys remove nitrogenous wastes from the blood and regulate ion and water levels in the body. (28)

Killer T Cells A type of lymphocyte that functions in the destruction of virus-infected cells and cancer cells. (30)

Kinases Enzymes that catalyze reactions in which phosphate groups are transferred from ATP to another molecule. (6)

Kinetic Energy Energy in motion. (6)

Kinetochore Part of a mitotic (or meiotic) chromosome that is situated within the centromere and to which the spindle fibers attach. (10)

Kingdom A level of the taxonomic hierarchy that groups together members of related phyla or divisions. Modern taxonomy divides all organisms into five Kingdoms: Monera, Protista, Fungi, Plantae, and Animalia. (1)

Klinefelter Syndrome A male whose cells have an extra X chromosome (XXY). The syndrome is characterized by underdeveloped male genitalia and feminine secondary sex characteristics. (17)

Krebs Cycle A circular pathway in aerobic respiration that completely oxidizes the two pyruvic acids from glycolysis. (9)

K-Selected Species Species that produce one or a few well-cared for individuals at a time. (43)

◄ L ►

Lacteal Blind lymphatic vessel in the intestinal villi that receives the absorbed products of lipid digestion. (27)

Lactic Acid Fermentation The process in which electrons removed during glycolysis are transferred from NADH to pyruvic acid to form lactic acid. Used by various prokaryotic cells under oxygen-deficient conditions and by muscle cells during strenuous activity. (9)

Lake Large body of standing fresh water, formed in natural depressions in the Earth. Lakes are larger than ponds. (40)

Lamella In bone, concentric cylinders of calcified collagen deposited by the osteocytes. The laminated layers produce a greatly strengthened structure. (26)

Large Intestine Portion of the intestine in which water and salts are reabsorbed. It is so named because of its large diameter. The large instestine, except for the rectum, is called the colon. (27)

Larva A self-feeding, sexually, and developmentally immature form of an animal. (32)

Larynx The short passageway connecting the pharynx with the lower airways. (29)

Latent (hidden) Infection Infection by a microorganism that causes no symptoms but the microbe is well-established in the body. (36)

Lateral Roots Roots that arise from the pericycle of older roots; also called branch roots or secondary roots. (18)

Law of Independent Assortment Alleles on nonhomologous chromosomes segregate independently of one another. (12)

Law of Segregation During gamete formation, pairs of alleles separate so that each sperm or egg cell has only one gene for a trait. (12)

Law of the Minimum The ecological principle that a species' distribution will be limited by whichever abiotic factor is most deficient in the environment. (41)

Laws of Thermodynamics Physical laws that describe the relationship of heat and mechanical energy. The first law states that energy cannot be created or destroyed, but one form can change into another. The second law states that the total energy the universe decreases as energy conversions occur and some energy is lost as heat. (6)

Leak Channels Passageways through a plasma membrane that do not contain gates and, therefore, are always open for the limited diffusion of a specific substance (ion) through the membrane. (7, 23)

Learning A process in which an animal benefits from experience so that its behavior is better suited to environmental conditions. (44)

Lenticels Loosely packed cells in the periderm of the stem that create air channels for transferring CO_2, H_2O, and O_2. (18)

Leukocytes White blood cells. (28)

LH Luteinizing hormone. A hormone secreted by the anterior pituitary that stimulates testosterone production in males and triggers ovulation and the transformation of the follicle into the corpus luteum in females. (31)

Lichen Symbiotic associations between certain fungi and algae. (37)

Life Cycle The sequence of events during the lifetime of an organism from zygote to reproduction. (39)

Ligaments Strong straps of connective tissue that hold together the bones in articulating joints or support an organ in place. (26)

Light-Dependent Reactions First stage of photosynthesis in which light energy is converted to chemical energy in the form of energy-rich ATP and NADPH. (8)

Light-Independent Reactions Second stage of photosynthesis in which the energy stored in ATP and NADPH formed in the light reactions is used to drive the reactions in which carbon dioxide is converted to carbohydrate. (8)

Limb Bud A portion of an embryo that will develop into either a forelimb or hindlimb. (32)

Limbic System A series of an interconnected group of brain structures, including the thalamus and hypothalamus, controlling memory and emotions. (23)

Limiting Factors The critical factors which impose restraints of the distribution, health, or activities of an organism. (41)

Limnetic Zone Open water of lakes, through which sunlight penetrates and photosynthesis occurs. (40)

Linkage The tendency of genes of the same chromosome to stay together rather than to assort independently. (13)

Linkage Groups Groups of genes located on the same chromosome. The genes of each linkage group assort independently of the genes of other linkage groups. In all eukaryotic organisms, the number of linkage groups is equal to the haploid number of chromosomes. (13)

Lipids A diverse group of biomolecules that are insoluble in water. (4)

Lithosphere The solid outer zone of the Earth; composed of the crust and outermost portion of the mantle. (40)

Littoral Zone Shallow, nutrient-rich waters of a lake, where sunlight reaches the bottom; also the lakeshore. Rooted vegetation occurs in this zone. (40)

Locomotion The movement of an organism from one place to another. (26)

Locus The chromosomal location of a gene. (13)

Logistic Growth Population growth producing a sigmoid, or S-shaped, growth curve. (43)

Long-Day Plants Plants that flower when the length of daylight exceeds some critical period. (21)

Longitudinal Fission The division pattern in flagellated protozoans, where division is along the length of the cell.

Loop of Henle An elongated section of the renal tubule that dips down into the kidney's medulla and then ascends back out to the cortex. It separates the proximal and distal convoluted tubules and is responsible for forming the salt gradient on which water reabsorption in the kidney depends. (28)

Low Density Lipoprotein (LDL) Particles that transport cholesterol in the blood. Each particle consists of about 1,500 cholesterol molecules surrounded by a film of phospholipids and protein. LDLs are taken into cells following their binding to cell surface LDL receptors. (7)

Low Intertidal Zone In the intertidal zone, the region which is uncovered by "minus" tides only. Organisms are submerged about 90% of the time. (40)

Lumen A space within an hollow organ or tube. (28)

Luminescence (see **Bioluminescence**)

Lungs The organs of terrestrial animals where gas exchange occurs. (29)

Lymph The colorless fluid in lymphatic vessels. (28)

Lymphatic System Network of fluid-carrying vessels and associated organs that participate in immunity and in the return of tissue fluid to the main circulation. (28)

Lymphocytes A group of non-phagocytic white blood cells which combat microbial invasion, fight cancer, and neutralize toxic chemicals. The two classes of lymphocytes, B cells and T cells, are the heart of the immune system. (28, 30)

Lymphoid Organs Organs associated with production of blood cells and the lymphatic system, including the thymus, spleen, appendix, bone marrow, and lymph nodes. (30)

Lysis (1) To split or dissolve. (2) Cell bursting.

Lysomes A type of storage vesicle produced by the Golgi complex, containing hydrolytic (digestive) enzymes capable of digesting many kinds of macromolecules in the cell. The membrane around them keeps them sequestered. (5)

◄ M ►

M Phase That portion of the cell cycle during which mitosis (nuclear division) and cytokinesis (cytoplasmic division) takes place. (10)

Macroevolution Evolutionary changes that lead to the appearance of new species. (33)

Macrofungus Filamentous fungus so named for the large size of its fleshy sexual structures; a mushroom, for example. (37)

Macromolecules Large polymers, such as proteins, nucleic acids, and polysaccharides. (4)

Macronutrients Nutrients required by plants in large amounts: carbon, oxygen, hydrogen, nitrogen, potassium, calcium, phosphorus, magnesium, and sulfur. (19)

Macrophages Phagocytic cells that develop from monocytes and present antigen to lymphocytes. (30)

Macroscopic Referring to biological observations made with the naked eye or a hand lens.

Mammals A class of vertebrates that possesses skin covered with hair and that nourishes their young with milk from mammary glands. (39)

Mammary Glands Glands contained in the breasts of mammalian mothers that produce breast milk. (39)

Marsupials Mammals with a cloaca whose young are born immature and complete their development in an external pouch in the mother's skin. (39)

Mass Extinction The simultaneous extinction of a multitude of species as the result of a drastic change in the environment. (33, 35)

Maternal Chromosomes The set of chromosomes in an individual that were inherited from the mother. (11)

Mechanoreceptors Sensory receptors that respond to mechanical pressure and detect motion, touch, pressure, and sound. (24)

Medulla The center-most portion of some organs. (23)

Medusa The motile, umbrella-shaped body form of some members of the phylum Cnidaria, with mouth and tentacles on the lower, concave service. (Compare with polyp.) (39)

Megaspores Spores that divide by mitosis to produce female gametophytes that produce the egg gamete. (20)

Meiosis The division process that produces cells with one-half the number of chromosomes in each somatic cell. Each resulting daughter cell is haploid (1N) (11)

Meiosis I A process of reductional division in which homologous chromosomes pair and then segregate. Homologues are partitioned into separate daughter cells. (11)

Meiosis II Second meiotic division. A division process resembling mitosis, except that the haploid number of chromosomes is present. After the chromosomes line up at the meta-phase plate, the two sister chromatids separate. (11)

Melanin A brown pigment that gives skin and hair its color (12)

Melanoma A deadly form of skin cancer that develops from pigment cells in the skin and is promoted by exposure to the sun. (14)

Memory Cells Lymphocytes responsible for active immunity. They recall a previous exposure to an antigen and, on subsequent exposure to the same antigen, proliferate rapidly into plasma cells and produce large quantities of antibodies in a short time. This protection typically lasts for many years. (30)

Mendelian Inheritance Transmission of genetic traits in a manner consistent with the principles discovered by Gregor Mendel. Includes traits controlled by simple dominant or recessive alleles; more complex patterns of transmission are referred to as Nonmendelian inheritance. (12)

Meninges The thick connective tissue sheath which surrounds and protects the vertebrate brain and spinal cord. (23)

Menstrual Cycle The repetitive monthly changes in the uterus that prepare the endometrium for receiving and supporting an embryo. (31)

Meristematic Region New cells arise from this undifferentiated plant tissue; found at root or shoot apical meristems, or lateral meristems. (18)

Meristems In plants, clusters of cells that retain their ability to divide, thereby producing new cells. One of the four basic tissues in plants. (18)

Mesoderm In animals, the middle germ cell layer of the gastrula. It gives rise to muscle, bone, connective tissue, gonads, and kidney. (32)

Mesopelagic Zone The dimly lit ocean zone beneath the epipelagic zone; large fishes, whales and squid occupy this zone; no phytoplankton occur in this zone. (40)

Mesophyll Layers of cells in a leaf between the upper and lower epidermis; produced by the ground meristem. (18)

Messenger RNA (mRNA) The RNA that carries genetic information from the DNA in the nucleus to the ribosomes in the cytoplasm, where the sequence of bases in the mRNA is translated into a sequence of amino acids. (14)

Metabolic Intermediates Compounds produced as a substrate are converted to end product in a series of enzymatic reactions. (6)

Metabolic Pathways Set of enzymatic reactions involved in either building or dismantling complex molecules. (6)

Metabolic Rate A measure of the level of activity of an organism usually determined by measuring the amount of oxygen consumed by an individual per gram body weight per hour. (22)

Metabolic Water Water produced as a product of metabolic reactions. (28)

Metabolism The sum of all the chemical reactions in an organism; includes all anabolic and catabolic reactions. (6)

Metamorphosis Transformation from one form into another form during development. (32)

Metaphase The stage of mitosis when the chromosomes line-up along the metaphase plate, a plate that usually lies midway between the spindle poles. (10)

Metaphase Plate Imaginary plane within a dividing cell in which the duplicated chromosomes become aligned during metaphase. (10)

Microbes Microscopic organisms. (36)

Microbiology The branch of biology that studies microorganisms. (36)

Microevolution Changes in allele frequency of a species' gene pool which has not generated new species. Exemplified by changes in the pigmentation of the peppered moth and by the acquisition of pesticide resistance in insects. (33)

Microfibrils Bundles formed from the intertwining of cellulose molecules, i.e., long chains of glucose molecules in the cell walls of plants. (5)

Microfilaments Thin actin-containing protein fibers that are responsible for maintenance of cell shape, muscle contraction and cyclosis. (5)

Micrometer One millionth (1/1,000,000) of a meter.

Micronutrients Nutrients required by plants in small amounts: iron, chlorine, copper, manganese, zinc, molybdenum, and boron. (19)

Micropyle A small opening in the integuments of the ovule through which the pollen tube grows to deliver sperm. (21)

Microspores Spores within anthers of flowers. They divide by mitosis to form pollen grains, the male gametophytes that produce the plant's sperm. (20)

Microtubules Thin, hollow tubes in cells; built from repeating protein units of tubulin. Microtubules are components of cilia, flagella, and the cytoskeleton. (5)

Microvilli The small projections on the cells that comprise each villus of the intestinal wall, further increasing the absorption surface area of the small intestine. (27)

Middle Intertidal Zone In the intertidal zone, the region which is covered and uncovered twice a day, the zero of tide tables. Organisms are submerged about 50% of the time. (40)

Migration Movements of a population into or out of an area. (44)

Mimicry A defense mechanism where one species resembles another in color, shape, behavior, or sound. (42)

Mineralocorticoids Steroid hormones which regulate the level of sodium and potassium in the blood. (25)

Mitochondria Organelles that contain the biochemical machinery for the Krebs cycle and the electron transport system of aerobic respiration. They are composed of two membranes, the inner one forming folds, or cristae. (9)

Mitosis The process of nuclear division producing daughter cells with exactly the same number of chromosomes as in the mother cell. (10)

Mitosis Promoting Factor (MPF) A protein that appears to be a universal trigger of cell division in eukaryotic cells. (10)

Mitotic Chromosomes Chromosomes whose DNA-protein threads have become coiled into microscopically visible chromosomes, each containing duplicated chromatids ready to be separated during mitosis. (10)

Molds Filamentous fungi that exist as colonies of threadlike cells but produce no macroscopic fruiting bodies. (37)

Molecule Chemical substance formed when two or more atoms bond together; the smallest unit of matter that possesses the qualities of a compound. (3)

Mollusca A phylum, second only to Arthropoda in diversity. Composed of three main classes: 1) Gastropoda (spiral-shelled), 2) Bivalvia (hinged shells), 3) Cephalopoda (with tentacles or arms and no, or very reduced shells). (39)

Molting (Ecdysis) Shedding process by which certain arthropods lose their exoskeletons as their bodies grow larger. (39)

Monera The taxonomic kingdom comprised of single-celled prokaryotes such as bacteria, cyanobacteria, and archebacteria. (36)

Monoclonal Antibodies Antibodies produced by a clone of hybridoma cells, all of which descended from one cell. (30)

Monocotyledae (Monocots) One of the two divisions of flowering plants, characterized by seeds with a single cotyledon, flower parts in 3s, parallel veins in leaves, many roots of approximately equal size, scattered vascular bundles in its stem anatomy, pith in its root anatomy, and no secondary growth capacity. (18)

Monocytes A type of leukocyte that gives rise to macrophages. (28)

Monoecious Both male and female reproductive structures are produced on the same sporophyte individual. (20, 38)

Monohybrid Cross A mating between two individuals that differ only in one genetically-determined trait. (12)

Monomers Small molecular subunits which are the building blocks of macromolecules. The macromolecules in living systems are constructed of some 40 different monomers. (4)

Monotremes A group of mammals that lay eggs from which the young are hatched. (39)

Morphogenesis The formation of form and internal architecture within the embryo brought about by such processes as programmed cell death, cell adhesion, and cell movement. (32)

Morphology The branch of biology that studies form and structure of organisms.

Mortality Deathrate in a population or area. (43)

Motile Capable of independent movement.

Motor Neurons Nerve cells which carry outgoing impulses to their effectors, either glands or muscles. (23)

Mucosa The cell layer that lines the digestive tract and secretes a lubricating layer of mucus. (27)

Mullerian Mimicry Resemblance of different species, each of which is equally obnoxious to predators. (42)

Multicellular Consisting of many cells. (35)

Multichannel Food Chain Where the same primary producer supplies the energy for more than one food chain. (41)

Multiple Allele System Three or more possible alleles for a given trait, such as ABO blood groups in humans. (12)

Multiple Fission Division of the cell's nucleus without a corresponding division of cytoplasm.

Multiple Fruits Fruits that develop from pistils of separate flowers. (20)

Muscle Fiber A muiltinucleated skeletal muscle cell that results from the fusion of several pre-muscle cells during embryonic development. (26)

Muscle Tissue Bundles and sheets of contractile cells that shorten when stimulated, providing force for controlled movement. (26)

Mutagens Chemical or physical agents that induce genetic change. (14)

Mutation Random heritable changes in DNA that introduce new alleles into the gene pool. (14)

Mutualism The symbiotic interaction in which both participants benefit. (42)

Mycology The branch of biology that studies fungi. (37)

Mycorrhizae An association between soil fungi and the roots of vascular plants, increasing the plant's ability to extract water and minerals from the soil. (19)

Myelin Sheath In vertebrates, a jacket which covers the axons of high-velocity neurons, thereby increasing the speed of a neurological impulse. (23)

Myofibrils In striated muscle, the banded fibrils that lie parallel to each other, constituting the bulk of the muscle fiber's interior and powering contraction. (26)

Myosin A contractile protein that makes up the major component of the thick filaments of a muscle cell and is also present in nonmuscle cells. (26)

◀ N ▶

NADPH Nicotinamide adenine dinucleotide phosphate. NADPH is formed by reduction of $NADP^+$, and serves as a store of electrons for use in metabolism (see Reducing Power). (9)

NAD^+ Nicotinamide adenine dinucleotide. A coenzyme that functions as an electron carrier in metabolic reactions. When reduced to NADH, the molecule becomes a cellular energy source. (9)

Natality Birthrate in a population or area. (43)

Natural Killer (NK) Cells Nonspecific, lymphocytelike cells which destroy foreign cells and cancer cells. (30)

Natural Selection Differential survival and reproduction of organisms with a resultant increase in the frequency of those best adapted to the environment. (33)

Neanderthals A subspecies of Homo sapiens different from that of modern humans that were characterized by heavy bony skeletons and thick bony ridges over the eyes. They disappeared about 35,000 years ago. (34)

Nectary Secretory gland in flowering plants containing sugary fluid that attracts pollinators as a food source. Usually located at the base of the flower. (20)

Negative Feedback Any regulatory mechanism in which the increased level of a substance inhibits further production of that substance, thereby preventing harmful accumulation. A type of homeostatic mechanism. (22, 25)

Negative Gravitropism In plants, growth against gravitational forces, or shoot growth upward. (21)

Nematocyst Within the stinging cell (cnidocyte) of cnidarians, a capsule that contains a coiled thread which, when triggered, harpoons prey and injects powerful toxins. (39)

Nematoda The widespread and abundant animal phylum containing the roundworms. (39)

Nephridium A tube surrounded by capillaries found in an organism's excretory organs that removes nitrogenous wastes and regulates the water and chemical balance of body fluids. (28)

Nephron The functional unit of the vertebrate kidney, consisting of the glomerulus, Bowman's capsule, proximal and distal convoluted tubules, and loop of Henle. (28)

Nerve Parallel bundles of neurons and their supporting cells. (23)

Nerve Impulse A propagated action potential. (23)

Nervous Tissue Excitable cells that receive stimuli and, in response, transmit an impulse to another part of the animal. (23)

Neural Plate In vertebrates, the flattened plate of dorsal ectoderm of the late gastrula that gives rise to the nervous system. (32)

Neuroglial Cells Those cells of a vertebrate nervous system that are not neurons. Includes a variety of cell types including Schwann cells. (23)

Neuron A nerve cell. (23)

Neurosecretory Cells Nervelike cells that secrete hormones rather than neurotransmitter substances when a nerve impulse reaches the distal end of the cell. In vertebrates, these cells arise from the hypothalamus. (25)

Neurotoxins Substances, such as curare and tetanus toxin, that interfere with the transmission of neural impulses. (23)

Neurotransmitters Chemicals released by neurons into the synaptic cleft, stimulating or inhibiting the post-synaptic target cell. (23)

Neurulation Formation by embryonic induction of the neural tube in a developing vertebrate embryo. (32)

Neutrons Electrically neutral (uncharged) particles contained within the nucleus of the atom. (3)

Neutrophil Phagocytic leukocyte, most numerous in the human body. (28)

Niche An organism's habitat, role, resource requirements, and tolerance ranges for each abiotic condition. (42)

Niche Breadth Relative size and dimension of ecological niches; for example, broad or narrow niches. (41)

Niche Overlap Organisms that have the same habitat, role, environmental requirements, or needs. (41)

Nitrogen Fixation The conversion of atmospheric nitrogen gas N_2 into ammonia (NH_3) by certain bacteria and cyanobacteria. (19)

Nitrogenous Wastes Nitrogen-containing metabolic waste products, such as ammonia or urea, that are produced by the breakdown of proteins and nucleic acids. (28)

Nodes The attachment points of leaves to a stem. (18)

Nodes of Ranvier Uninsulated (nonmyelinated) gaps along the axon of a neuron. (23)

Noncovalent Bonds Linkages between two atoms that depend on an attraction between positive and negative charges between molecules or ions. Includes ionic and hydrogen bonds. (3)

Non-Cyclic Photophosphorylation The pathway in the light reactions of photosynthesis in which electrons pass from water, through two photosystems, and then ultimately to $NADP^+$. During the process, both ATP and NADPH are produced. It is so named because the electrons do not return to their reaction center. (8)

Nondisjunction Failure of chromosomes to separate properly at meiosis I or II. The result is that one daughter will receive an extra chromosome and the other gets one less. (11, 13)

Nonpolar Molecules Molecules which have an equal charge distribution throughout their structure and thus lack regions with a localized positive or negative charge. (3)

Notochord A flexible rod that is below the dorsal surface of the chordate embryo, beneath the nerve cord. In most chordates, it is replaced by the vertebral column. (32)

Nuclear Envelope A double membrane pierced by pores that separates the contents of the nucleus from the rest of the eukaryotic cell. (5)

Nucleic Acids DNA and RNA; linear polymers of nucleotides, responsible for the storage and expression of genetic information. (4, 14)

Nucleoid A region in the prokaryotic cell that contains the genetic material (DNA). It is unbounded by a nuclear membrane. (36)

Nucleoplasm The semifluid substance of the nucleus in which the particulate structures are suspended. (5)

Nucleosomes Nuclear protein complex consisting of a length of DNA wrapped around a central cluster of 8 histones. (14)

Nucleotides Monomers of which DNA and RNA are built. Each consists of a 5-carbon sugar, phosphate, and a nitrogenous base. (4)

Nucleous (pl. nucleoli) One or more darker regions of a nucleus where each ribosomal subunit is assembled from RNA and protein. (5)

Nucleus The large membrane-enclosed organelle that contains the DNA of eukaryotic cells. (5)

Nucleus, Atomic The center of an atom containing protons and neutrons. (3)

◄ O ►

Obligate Symbiosis A symbiotic relationship between two organisms that is necessary for the survival or both organisms. (42)

Olfaction The sense of smell. (24)

Oligotrophic Little nourished, as a young lake that has few nutrients and supports little life. (40)

Omnivore An animal that obtains its nutritional needs by consuming plants and other animals. (42)

Oncogene A gene that causes cancer, perhaps activated by mutation or a change in its chromosomal location. (10)

Oocyte A female germ cell during any of the stages of meiosis. (31)

Oogenesis The process of egg production. (31)

Oogonia Female germ cells that have not yet begun meiosis. (31)

Open Circulatory System Circulatory system in which blood travels from vessels to tissue spaces, through which it percolates prior to returning to the main vessel (compare with closed circulatory system). (28)

Operator A regulatory gene in the operon of bacteria. It is the short DNA segment to which the repressor binds, thus preventing RNA polymerase from attaching to the promoter. (15)

Operon A regulatory unit in prokaryotic cells that controls the expression of structural genes. The operon consists of structural genes that produce enzymes for a particular metabolic pathway, a regulator region composed of a promoter and an operator, and R (regulator) gene that produces a repressor. (15)

Order A level of the taxonomic hierarchy that groups together members of related families. (1)

Organ Body part composed of several tissues that performs specialized functions. (22)

Organelle A specialized part of a cell having some particular function. (5)

Organic Compounds Chemical compounds that contain carbon. (4)

Organism A living entity able to maintain its organization, obtain and use energy, reproduce, grow, respond to stimuli, and display homeostatis. (1)

Organogenesis Organ formation in which two or more specialized tissue types develop in a precise temporal and spatial relationship to each other. (32)

Organ System Group of functionally related organs. (22)

Osmoregulation The maintenance of the proper salt and water balance in the body's fluids. (28)

Osmosis The diffusion of water through a differentially permeable membrane into a hypertonic compartment. (7)

Ossification Synthesis of a new bone. (26)

Osteoclast A type of bone cell which breaks down the bone, thereby releasing calcium into the bloodstream for use by the body. Osteoclasts are activated by hormones released by the parathyroid glands. (26)

Osteocytes Living bone cells embedded within the calcified matrix they manufacture. (26)

Osteoporosis A condition present predominantly in postmenopausal women where the bones are weakened due to an increased rate of bone resorption compared to bone formation. (26)

Ovarian Cycle The cycle of egg production within the mammalian ovary. (31)

Ovarian Follicle In a mammalian ovary, a chamber of cells in which the oocyte develops. (31)

Ovary In animals, the egg-producing gonad of the female. In flowering plants, the enlarged base of the pistil, in which seeds develop. (20)

Oviduct (Fallopian Tube) The tube in the female reproductive organ that connects the ovaries and uterus and where fertilization takes place. (31)

Ovulation The release of an egg (ovum) from the ovarian follicle. (31)

Ovule In seed plants, the structure containing the female gametophyte, nucellus, and integuments. After fertilization, the ovule develops into a seed. (20, 38)

Ovum An unfertilized egg cell; a female gamete. (31)

Oxidation The removal of electrons from a compound during a chemical reaction. For a carbon atom, the fewer hydrogens bonded to a carbon, the greater the oxidation state of the atom. (6)

Oxidative Phosphorylation The formation of ATP from ADP and inorganic phosphate that occurs in the electron-transport chain of cellular respiration. (8, 9)

Oxyhemoglobin A complex of oxygen and hemoglobin, formed when blood passes through the lungs and is dissociated in body tissues, where oxygen is released. (29)

Oxytocin A female hormone released by the posterior pituitary which triggers uterine contractions during childbirth and the release of milk during nursing. (25)

◄ P ►

P680 Reaction Center (P = Pigment) Special chlorophyll molecule in Photosystem II that traps the energy absorbed by the other pigment molecules. It absorbs light energy maximally at 680 nm. (8)

Palisade Parenchyma In dicot leaves, densely packed, columnar shaped cells functioning in photosynthesis. Found just beneath the upper epidermis. (18)

Pancreas In vertebrates, a large gland that produces digestive enzymes and hormones. (27)

Parallel Evolution When two species that have descended from the same ancestor independently acquire the same evolutionary adaptations. (33)

Parapatric Speciation The splitting of a population into two species' populations under conditions where the members of each population reside in adjacent areas. (33)

Parasite An organism that lives on or inside another called a host, on which it feeds. (39, 42)

Parasitism A relationship between two organisms where one organism benefits, and the other is harmed. (42)

Parasitoid Parasitic organisms, such as some insect larvae, which kill their host. (42)

Parasympathetic Nervous System Part of the autonomic nervous system active during relaxed activity. (23)

Parathyroid Glands Four glands attached to the thyroid gland which secrete parathyroid hormone (PTH). When blood calcium levels are low, PTH is secreted, causing calcium to be released from bone. (25)

Parenchyma The most prevalent cell type in herbaceous plants. These thin-walled, polygonal-shaped cells function in photosynthesis and storage. (18)

Parthenogenesis Process by which offspring are produced without egg fertilization. (31)

Passive Immunity Immunity achieved by receiving antibodies from another source, as occurs with a newborn infant during nursing. (30)

Paternal Chromosomes The set of chromosomes in an individual that were inherited from the father. (11)

Pathogen A disease-causing microorganism. (36)

Pectoral Girdle In humans, the two scapulae (shoulder blades) and two clavicles (collarbones) which support and articulate with the bones of the upper arm. (26)

Pedicel A shortened stem carrying a flower. (20)

Pedigree A diagram showing the inheritance of a particular trait among the members of the family. (13)

Pelagic Zone The open oceans, divided into three layers: 1) photo- or epipelagic (sunlit), 2) mesopelagic (dim light), 3) aphotic or bathypelagic (always dark). (40)

Pelvic Girdle The complex of bones that connect a vertebrate's legs with its backbone. (26)

Penis An intrusive structure in the male animal which releases male gametes into the female's sex receptacle. (31)

Peptide Bond The covalent bond between the amino group of one amino acid and the carboxyl group of another. (4)

Peptidoglycan A chemical component of the prokaryotic cell wall. (36)

Percent Annual Increase A measure of population increase; the number of individuals (people) added to the population per 100 individuals. (43)

Perennials Plants that live longer than two years. (18)

Perfect Flower Flowers that contain both stamens and pistils. (20)

Perforation Plate In plants, that portion of the wall of vessel members that is perforated, and contains an area with neither primary nor secondary cell wall; a "hole" in the cell wall. (18)

Pericycle One or more layers of cells found in roots, with phloem or xylem to its' inside, and the endodermis to its' outside. Functions in producing lateral roots and formation of the vascular cambium in roots with secondary growth. (18)

Periderm Secondary tissue that replaces the epidermis of stems and roots. Consists of cork, cork cambium, and an internal layer of parenchyma cells. (18)

Peripheral Nervous System Neurons, excluding those of the brain and spinal cord, that permeate the rest of the body. (23)

Peristalsis Sequential waves of muscle contractions that propel a substance through a tube. (27)

Peritoneum The connective tissue that lines the coelomic cavities. (39)

Permeability The ability to be penetrable, such as a membrane allowing molecules to pass freely across it. (7)

Petal The second whorl of a flower, often brightly colored to attract pollinators; collectively called the corolla. (20)

Petiole The stalk leading to the blade of a leaf. (18)

pH A scale that measures the concentration of hydrogen ions in a solution. The pH scale extends from 0 to 14. Acidic solutions have a pH of less than 7; alkaline solutions have a pH above 7; neutral solutions have a pH equal to 7. (3)

Phagocytosis Engulfing of food particles and foreign cells by amoebae and white blood cells. A type of endocytosis. (5)

Pharyngeal Pouches In the vertebrate embryo, outgrowths from the walls of the pharynx that break through the body surface to form gill slits. (32)

Pharynx The throat; a portion of both the digestive and respiratory system just behind the oral cavity. (29)

Phenotype An individual's observable characteristics that are the expression of its genotype. (12)

Pheromones Chemicals that, when released by an animal, elicit a particular behavior in other animals of the same species. (44)

Phloem The vascular tissue that transports sugars and other organic molecules from sites of photosynthesis and storage to the rest of the plant. (18)

Phloem Loading The transfer of assimilates to phloem conducting cells, from photosynthesizing source cells. (19)

Phloem Unloading The transfer of assimilates to storage (sink) cells, from phloem conducting cells. (19)

Phospholipids Lipids that contain a phosphate and a variable organic group that form polar, hydrophilic regions on an otherwise nonpolar, hydrophobic molecule. They are the major structural components of membranes. (4)

Phosphorylation A chemical reaction in which a phosphate group is added to a molecule or atom. (6)

Photoexcitation Absorption of light energy by pigments, causing their electrons to be raised to a higher energy level. (8)

Photolysis The splitting of water during photosynthesis. The electrons from water pass to Photosystem II, the protons enter the lumen of the thylakoid and contribute to the proton gradient across the thylakoid membrane, and the oxygen is released into the atmosphere. (8)

Photon A particle of light energy. (8)

Photoperiod Specific lengths of day and night which control certain plant growth responses to light, such as flowering or germination. (21)

Photoperiodism Changes in the behavior and physiology of an organism in response to the relative lengths of daylight and darkness, i.e., the photoperiod. (21)

Photoreceptors Sensory receptors that respond to light. (24)

Photorespiration The phenomenon in which oxygen binds to the active site of a CO_2-fixing enzyme, thereby competing with CO_2 fixation, and lowering the rate of photosynthesis. (8)

Photosynthesis The conversion by plants of light energy into chemical energy stored in carbohydrate. (8)

Photosystems Highly organized clusters of photosynthetic pigments and electron/hydrogen carriers embedded in the thylakoid membranes of chloroplasts. There are two photosystems, which together carry out the light reactions of photosynthesis. (8)

Photosystem I Photosystem with a P700 reaction center; participates in cyclic photophosphorylation as well as in noncyclic photophosphorylation. (8)

Photosystem II Photosystem activated by a P680 reaction center; participates only in noncyclic photophosphorylation and is associated with photolysis of water. (8)

Phototropism The growth responses of a plant to light. (21)

Phyletic Evolution The gradual evolution of one species into another. (33)

Phylogeny Evolutionary history of a species. (35)

Phylum The major taxonomic divisions in the Animal kingdom. Members of a phylum share common, basic features. The Animal kingdom is divided into approximately 35 phyla. (39)

Physiology The branch of biology that studies how living things function. (22)

Phytochrome A light-absorbing pigment in plants which controls many plant responses, including photoperiodism. (21)

Phytoplankton Microscopic photosynthesizers that live near the surface of seas and bodies of fresh water. (37)

Pineal Gland An endocrine gland embedded within the brain that secretes the hormone melatonin. Hormone secretion is dependent on levels of environmental light. In amphibians and reptiles, melatonin controls skin coloration. In humans, pineal secretions control sexual maturation and daily rhythms. (25)

Pinocytosis Uptake of small droplets and dissolved solutes by cells. A type of endocytosis. (5)

Pistil The female reproductive part and central portion of a flower, consisting of the ovary, style and stigma. May contain one carpel, or one or more fused carpels. (20)

Pith A plant tissue composed of parenchyma cells, found in the central portion of primary growth stems of dicots, and monocot roots. (18)

Pith Ray Region between vascular bundles in vascular plants. (18)

Pituitary Gland (see **Posterior and Anterior Pituitary).**

Placenta In mammals (exclusive of marsupials and monotremes), the structure through which nutrients and wastes are exchanged between the mother and embryo/fetus. Develops from both embryonic and uterine tissues. (32)

Plant Multicellular, autotrophic organism able to manufacture food through photosynthesis. (38)

Plasma In vertebrates, the liquid portion of the blood, containing water, proteins (including fibrinogen), salts, and nutrients. (28)

Plasma Cells Differentiated antibody-secreting cells derived from B lymphocytes. (30)

Plasma Membrane The selectively permeable, molecular boundary that separates the cytoplasm of a cell from the external environment. (5)

Plasmid A small circle of DNA in bacteria in addition to its own chromosome. (16)

Plasmodesmata Openings between plant cell walls, through which adjacent cells are connected via cytoplasmic threads. (19)

Plasmodium Genus of protozoa that causes malaria. (37)

Plasmodium A huge multinucleated "cell" stage of a plasmodial slime mold that feeds on dead organic matter. (37)

Plasmolysis The shrinking of a plant cell away from its cell wall when the cell is placed in a hypertonic solution. (7)

Platelets Small, cell-like fragments derived from special white blood cells. They function in clotting. (28)

Plate Tectonics The theory that the earth's crust consists of a number of rigid plates that rest on an underlying layer of semimolten rock. The movement of the earth's plates results from the upward movement of molten rock into the solidified crust along ridges within the ocean floor. (35)

Platyhelminthes The phylum containing simple, bilaterally symmetrical animals, the flatworms. (39)

Pleiotropy Where a single mutant gene produces two or more phenotypic effects. (12)

Pleura The double-layered sac which surrounds the lungs of a mammal. (29, 39)

Pneumocytis Pneumonia (PCP) A disease of the respiratory tract caused by a protozoan that strikes persons with immunodeficiency diseases, such as AIDS. (30)

Point Mutations Changes that occur at one point within a gene, often involving one nucleotide in the DNA. (14)

Polar Body A haploid product of meiosis of a female germ cell that has very little cytoplasm and disintegrates without further function. (31)

Polar Molecule A molecule with an unequal charge distribution that creates distinct positive and negative regions or poles. (3)

Pollen The male gametophyte of seed plants, comprised of a generative nucleus and a tube nucleus surrounded by a tough wall. (20)

Pollen Grain The male gametophyte of conifers and angiosperms, containing male gametes. In angiosperms, pollen grains are contained in the pollen sacs of the anther of a flower. (20)

Pollination The transfer of pollen grains from the anther of one flower to the stigma of another. The transfer is mediated by wind, water, insects, and other animals. (20)

Polygenic Inheritance An inheritance pattern in which a phenotype is determined by two or more genes at different loci. In humans, examples include height and pigmentation. (12)

Polymer A macromolecule formed of monomers joined by covalent bonds.. Includes proteins, polysaccharides, and nucleic acids. (4)

Polymerase Chain Reaction (PCR) Technique to amplify a specific DNA molecule using a temperature-sensitive DNA polymerase obtained from a heat-resistant bacterium. Large numbers of copies of the initial DNA molecule can be obtained in a short period of time, even when the starting material is present in vanishingly small amounts, as for example from a blood stain left at the scene of a crime. (16)

Polymorphic Property of some protozoa to produce more than one stage of organism as they complete their life cycles. (37)

Polymorphic Genes Genes for which several different alleles are known, such as those that code for human blood type. (17)

Polyp Stationary body form of some members of the phylum Cnidaria, with mouth and tentacles facing upward. (Compare with medusa.) (39)

Polypeptide An unbranched chain of amino acids covalently linked together and assembled on a ribosome during translation. (4)

Polyploidy An organism or cell containing three or more complete sets of chromosomes. Polyploidy is rare in animals but common in plants. (33)

Polysaccharide A carbohydrate molecule consisting of monosaccharide units. (4)

Polysome A complex of ribosomes found in chains, linked by mRNA. Polysomes contain the ribosomes that are actively assembling proteins. (14)

Polytene Chromosomes Giant banded chromosomes found in certain insects that form by the repeated duplication of DNA. Because of the multiple copies of each gene in a cell, polytene chromosomes can generate large amounts of a gene product in a short time. Transcription occurs at sites of chromosome puffs. (13)

Pond Body of standing fresh water, formed in natural depressions in the Earth. Ponds are smaller than lakes. (40)

Population Individuals of the same species inhabiting the same area. (43)

Population Density The number of individual species living in a given area. (43)

Positive Gravitropism In plants, growth with gravitational forces, or root growth downward. (21)

Posterior Pituitary A gland which manufactures no hormones but receives and later releases hormones produced by the cell bodies of neurons in the hyopthalamus. (25)

Potential Energy Stored energy, such as occurs in chemical bonds. (6)

Preadaptation A characteristic (adaptation) that evolved to meet the needs of an organism in one type of habitat, but fortuitously allows the organism to exploit a new habitat. For example, lobed fins and lungs evolved in ancient fishes to help them live in shallow, stagnant ponds, but also facilitated the evolution of terrestrial amphibians. (33, 39)

Precells Simple forerunners of cells that, presumably, were able to concentrate organic molecules, allowing for more frequent molecular reactions. (35)

Predation Ingestion of prey by a predator for energy and nutrients. (42)

Predator An organism that captures and feeds on another organism (prey). (42)

Pressure Flow In the process of phloem loading and unloading, pressure differences resulting from solute increases in phloem conducting cells and neighboring xylem cells cause the flow of water to phloem. A concentration gradient is created between xylem and phloem cells. (19)

Prey An organism that is captured and eaten by another organism (predator). (42)

Primary Consumer Organism that feeds exclusively on producers (plants). Herbivores are primary consumers. (41)

Primary Follicle In the mammalian ovary, a structure composed of an oocyte and its surrounding layer of follicle cells. (31)

Primary Growth Growth from apical meristems, resulting in an increase in the lengths of shoots and roots in plants. (18)

Primary Immune Response Process of antibody production following the first exposure to an antigen. There is a lag time from exposure until the appearance in the blood of protective levels of antibodies. (30)

Primary Oocyte Female germ cell that is either in the process of or has completed the first meiotic division. In humans, germ cells may remain in this stage in the ovary for decades. (31)

Primary Producers All autotrophs in a biotic environment that use sunlight or chemical energy to manufacture food from inorganic substances. (41)

Primary Sexual Characteristics Gonads, reproductive tracts, and external genitals. (31)

Primary Spermatocyte Male germ cell that is either in the process of or has completed the first meiotic division. (31)

Primary Succession The development of a community in an area previously unoccupied by any community; for example, a "bare" area such as rock, volcanic material, or dunes. (41)

Primary Tissues Tissues produced by primary meristems of a plant, which arise from the shoot and root apical meristems. In general, primary tissues are a result of an increase in plant length. (18)

Primary Transcript An RNA molecule that has been transcribed but not yet subjected to any type of processing. The primary transcript corresponds to the entire stretch of DNA that was transcribed. (15)

Primates Order of mammals that includes humans, apes, monkeys, and lemurs. (39)

Primitive An evolutionary early condition. Primitive features are those that were also present in an early ancestor, such as five digits on the feet of terrestrial vertebrates. (34)

Prions An infectious particle that contains protein but no nucleic acid. It causes slow diseases of animals, including neurological disease of humans. (36)

Processing-Level Control Control of gene expression by regulating the pathway by which a primary RNA transcript is processed into an mRNA. (15)

Products In a chemical reaction, the compounds into which the reactants are transformed. (6)

Profundal Zone Deep, open water of lakes, where it is too dark for photosynthesis to occur. (40)

Progesterone A hormone produced by the corpus luteum within the ovary. It prepares and maintains the uterus for pregnancy, participates in milk production, and prevents the ovary from releasing additional eggs late in the cycle or during pregnancy. (25)

Prokaryotic Referring to single-celled organisms that have no membrane separating the DNA from the cytoplasm and lack membrane-enclosed organelles. Prokaryotes are confined to the kingdom Monera; they are all bacteria. (36)

Prokaryotic Fission The most common type of cell division in bacteria (prokaryotes). Duplicated DNA strands are attached to the plasma membrane and become separated into two cells following membrane growth and cell wall formation. (10, 36)

Prolactin A hormone produced by the anterior pituitary, stimulating milk production by mammary glands. (25)

Promoter A short segment of DNA to which RNA polymerase attaches at the start of transcription. (15)

Prophase Longest phase of mitosis, involving the formation of a spindle, coiling of chromatin fibers into condensed chromosomes, and movement of the chromosomes to the center of the cell. (10)

Prostaglandins Hormones secreted by endocrine cells scattered throughout the body responsible for such diverse functions as contraction of uterine muscles, triggering the inflammatory response, and blood clotting. (25)

Prostate Gland A muscular gland which produces and releases fluids that make up a substantial portion of the semen. (31)

Proteins Long chains of amino acids, linked together by peptide bonds. They are folded into specific shapes essential to their functions. (4)

Prothallus The small, heart-shaped gametophyte of a fern. (38)

Protists A member of the kingdom Protista; simple eukaryotic organisms that share broad taxonomic similarities. (36, 37)

Protocooperation Non-compulsory interactions that benefit two organisms, e.g., lichens. (42)

Proton Gradient A difference in hydrogen ion (proton) concentration on opposite sides of a membrane. Proton gradients are formed during photosynthesis and respiration and serve as a store of energy used to drive ATP formation. (8, 9)

Protons Positively charged particles within the nucleus of an atom. (3)

Protostomes One path of development exhibited by coelomate animals (e.g., mollusks, annelids, and arthropods). (39)

Protozoa Member of protist kingdom that is unicellular and eukaryotic; vary greatly in size, motility, nutrition and life cycle. (37)

Provirus DNA copy of a virus' nucleic acid that becomes integrated into the host cell's chromosome. (36)

Pseudocoelamates Animals in which the body cavity is not lined by cells derived from mesoderm. (39)

Pseudopodia (psuedo = false, pod = foot). Pseudopodia are fingerlike extensions of cytoplasm that flow forward from the "body" of an amoeba; the rest of the cell then follows. (37)

Puberty Development of reproductive capacity, often accompanied by the appearance of secondary sexual characteristics. (31)

Pulmonary Circulation The loop of the circulatory system that channels blood to the lungs for oxygenation. (28)

Punctuated Equilibrium Theory A theory to explain the phenomenon of the relatively sudden appearance of new species, followed by long periods of little or no change. (33)

Punnett Square Method A visual method for predicting the possible genotypes and their expected ratios from a cross. (12)

Pupa In insects, the stage in metamorphosis between the larva and the adult. Within the pupal case, there is dramatic transformation in body form as some larval tissues die and others differentiate into those of the adult. (32)

Purine A nitrogenous base found in DNA and RNA having a double ring structure. Adenine and guanine are purines. (14)

Pyloric Sphincter Muscular valve between the human stomach and small intestine. (27)

Pyrimidine A nitrogenous base found in DNA and RNA having a single ring structure. Cytosine, thymine, and uracil are pyrimidines. (14)

Pyramid of Biomass Diagrammatic representation of the total dry weight of organisms at each trophic level in a food chain or food web. (41)

Pyramid of Energy Diagrammatic representation of the flow of energy through trophic levels in a food chain or food web. (41)

Pyramid of Numbers Similar to a pyramid of energy, but with numbers of producers and consumers given at each trophic level in a food chain or food web. (41)

◄ Q ►

Quiescent Center The region in the apical meristem of a root containing relatively inactive cells. (18)

◀ R ▶

R-Group The variable portion of a molecule. (4)

r-Selected Species Species that possess adaptive strategies to produce numerous offspring at once. (43)

Radial Symmetry The quality possessed by animals whose bodies can be divided into mirror images by more than one median plane. (39)

Radicle In the plant embryo, the tip of the hypocotyl that eventually develops into the root system. (20)

Radioactivity A property of atoms whose nucleus contains an unstable combination of particles. Breakdown of the nucleus causes the emission of particles and a resulting change in structure of the atom. Biologists use this property to track labeled molecules and to determine the age of fossils. (3)

Radiodating The use of known rates of radioactive decay to date a fossil or other ancient object. (3, 34)

Radioisotope An isotope of an element that is radioactive. (3)

Radiolarian A prozoan member of the protistan group Sarcodina that secretes silicon shells through which it captures food.

Rainshadow The arid, leeward (downwind) side of a mountain range. (40)

Random Distribution Distribution of individuals of a population in a random manner; environmental conditions must be similar and individuals do not affect each other's location in the population. (43)

Reactants Molecules or atoms that are changed to products during a chemical reaction. (6)

Reaction A chemical change in which starting molecules (reactants) are transformed into new molecules (products). (6)

Reaction Center A special chlorophyll molecule in a photosystem (P_{700} in Photosystem I, P_{680} in Photosystem II). (8)

Realized Niche Part of the fundamental niche of an organism that is actually utilized. (41)

Receptacle The base of a flower where the flower parts are attached; usually a widened area of the pedicel. (20)

Receptor-Mediated Endocytosis The uptake of materials within a cytoplasmic vesicle (endocytosis) following their binding to a cell surface receptor. (7)

Receptor Site A site on a cell's plasma membrane to which a chemical such as a hormone binds. Each surface site permits the attachment of only one kind of hormone. (5)

Recessive An allele whose expression is masked by the dominant allele for the same trait. (12)

Recombinant DNA A DNA molecule that contains DNA sequences derived from different biological sources that have been joined together in the laboratory. (16)

Recombination The rejoining of DNA pieces with those of a different strand or with the same strand at a point different from where the break occurred. (11, 13)

Red Marrow The soft tissue in the interior of bones that produces red blood cells. (26)

Red Tide Growth of one of several species of reddish brown dinoflagellate algae so extensive that it tints the coastal waters and inland lakes a distinctive red color. Often associated with paralytic shellfish poisoning (see dinoflagellates). (37)

Reducing Power A measure of the cell's ability to transfer electrons to substrates to create molecules of higher energy content. Usually determined by the available store of NADPH, the molecule from which electrons are transferred in anabolic (synthetic) pathways. (6)

Reduction The addition of electrons to a compound during a chemical reaction. For a carbon atom, the more hydrogens that are bonded to the carbon, the more reduced the atom. (6)

Reduction Division The first meiotic division during which a cell's chromosome number is reduced in half. (11)

Reflex An involuntary response to a stimulus. (23)

Reflex Arc The simplest example of central nervous system control, involving a sensory neuron, motor neuron, and usually an interneuron. (23)

Regeneration Ability of certain animals to replace injured or lost limbs parts by growth and differentiation of undifferentiated stem cells. (15)

Region of Elongation In root tips, the region just above the region of cell division, where cells elongate and the root length increases. (18)

Region of Maturation In root tips, the region above the region of elongation; cells differentiate and root hairs occur in this region. (18)

Regulatory Genes Genes whose sole function is to control the expression of structural genes. (15)

Releaser A sign stimulus that is given by an individual to another member of the same species, eliciting a specific innate behavior. (44)

Releasing Factors Hormones secreted by the tips of hypothalmic neurosecretory cells that stimulate the anterior pituitary to release its hormones. GnRH, for example, stimulates the release of gonadotropins. (25)

Renal Referring to the kidney. (28)

Replication Duplication of DNA, usually prior to cell division. (14)

Replication Fork The site where the two strands of a DNA helix are unwinding during replication. (14)

Repression Inhibition of transcription of a gene which, in an operon, occurs when repressor protein binds to the operator. (15)

Repressor Protein encoded by a bacterial regulatory gene that binds to an operator site of an operon and inhibits transcription. (15)

Reproduction The process by which an organism produces offspring. (31)

Reproductive Isolation Phenomenon in which members of a single population become split into two populations that no longer interbreed. (33)

Reproductive System System of specialized organs that are utilized for the production of gametes and, in some cases, the fertilization and/or development of an egg. (31)

Reptiles Members of class Reptilia, scaly, air-breathing, egg-laying vertebrates such as lizards, snakes, turtles, and crocodiles. (39)

Resolving Power The ability of an optical instrument (eye, microscopes) to discern whether two very close objects are separate from each other. (APP.)

Resource Partitioning Temporal or spatial sharing of a resource by different species. (42)

Respiration Process used by organisms to exchange gases with the environment; the source of oxygen required for metabolism. The process organisms use to oxidize glucose to CO_2 and H_2O using an electron transport system to extract energy from electrons and store it in the high-energy bonds of ATP. (29)

Respiratory System The specialized set of organs that function in the uptake of oxygen from the environment. (29)

Resting Potential The electrical potential (voltage) across the plasma membrane of a neuron when the cell is not carrying an impulse. Results from a difference in charge across the membrane. (23)

Restriction Enzyme A DNA-cutting enzyme found in bacteria. (16)

Restriction Fragment Length Polymorphism (RFLP) Certain sites in the DNA tend to have a highly variable sequence from one individual to another. Because of these differences, restriction enzymes cut the DNA from different individuals into fragments of different length. Variations in the length of particular fragments (RFLPs) can be used as genetic signposts for the identification of nearby genes of interest. (17)

Restriction Fragments The DNA fragments generated when purified DNA is treated with a particular restriction enzyme. (16)

Reticular Formation A series of interconnected sites in the core of the brain (brainstem) that selectively arouse conscious activity. (23)

Retroviruses RNA viruses that reverse the typical flow of genetic information; within the infected cell, the viral DNA serves as a template for synthesis of a DNA copy. Examples include HIV, which causes AIDS, and certain cancer viruses. (36)

Reverse Genetics Determining the amino acid sequence and function of a polypeptide from the nucleotide sequence of the gene that codes for that polypeptide. (17)

Reverse Transcriptase An enzyme present in retroviruses that transcriibes a strand of DNA, using viral RNA as the template. (36)

Rhizoids Slender cells that resemble roots but do not absorb water or minerals. (36)

Rhodophyta Red algae; seaweeds that can absorb deeper penetrating light rays than most aquatic photosynthesizers. (36)

Rhyniophytes Ancient plants having vascular tissue which thrived in marshy areas during the Silurian period.

Ribonucleic Acid (RNA) Single-stranded chain of nucleotides each comprised of ribose (a sugar), phosphate, and one of four bases (adenine, guanine, cytosine, and uracil). The sequence of nucleotides in RNA is dictated by DNA, from which it is transcribed. There are three classes of RNA: mRNA, tRNA, and rRNA, all required for protein synthesis. (4, 14)

Ribosomal RNA (rRNA) RNA molecules that form part of the ribosome. Included among the rRNAs is one that is thought to catalyze peptide bond formation. (14)

Ribosomes Organelles involved in protein synthesis in the cytoplasm of the cell. (14)

Ribozymes RNAs capable of catalyzing a chemical reaction, such as peptide bond formation or RNA cutting and splicing. (15)

Rickettsias A group of obligate intracellular parasites, smaller than the typical prokaryote. They cause serious diseases such as typhus. (36)

River Flowing body of surface fresh water; rivers are formed from the convergence of streams. (40)

RNA Polymerase The enzyme that directs transcription and assembling RNA nucleotides in the growing chain. (14)

RNA Processing The process by which the intervening (noncoding) portions of a primary RNA transcript are removed and the remaining (coding) portions are spliced together to form an mRNA. (15)

Root Cap A protective cellular helmet at the tip of a root that surrounds delicate meristematic cells and shields them from abrasion and directs the growth downward. (18)

Root Hairs Elongated surface cells near the tip of each root for the absorption of water and minerals. (18)

Root Nodules Knobby structures on the roots of certain plants. They house nitrogen-fixing bacteria which supply nitrogen in a form that can be used by the plant. (19)

Root Pressure A positive pressure as a result of continuous water supply to plant roots that assists (along with transpirational pull) the pushing of water and nutrients up through the xylem. (19)

Root System The below-ground portion of a plant, consisting of main roots, lateral roots, root hairs, and associated structures and systems such as root nodules or mycorrhizae. (18)

Rough ER (RER) Endoplasmatic reticulum with many ribosomes attached. As a result, they appear rough in electron micrographs. (5)

Ruminant Grazing mammals that possess an additional stomach chamber called rumen which is heavily fortified with cellulose-digesting microorganisms. (27)

◄ S ►

S Phase The second stage of interphase in which the materials needed for cell division are synthesized and an exact copy of cell's DNA is made by DNA replication. (10)

Sac Body The body plan of simple animals, like cnidarians, where there is a single opening leading to and from a digestion chamber.

Saltatory Conduction The "hopping" movement of an impulse along a myelinated neuron from one Node of Ranvier to the next one. (23)

Sap Fluid found in xylem or sieve of phloem. (20)

Saprophyte Organisms, mainly fungi and bacteria, that get their nutrition by breaking down organic wastes and dead organisms, also called decomposers. (42)

Saprobe Organism that obtains its nutrients by decomposing dead organisms. (37)

Sarcolemma The plasma membrane of a muscle fiber. (26)

Sarcomere The contractile unit of a myofibril in skeletal muscle. (26)

Sarcoplasmic Reticulum (SR) In skeletal muscle, modified version of the endoplasmic reticulum that stores calcium ions. (26)

Savanna A grassland biome with alternating dry and rainy seasons. The grasses and scattered trees support large numbers of grazing animals. (40)

Scaling Effect A property that changes disproportionately as the size of organisms increase. (22)

Scanning Electron Microscope (SEM) A microscope which operates by showering electrons back and forth across the surface of a specimen prepared with a thin metal coating. The resultant image shows three-dimensional features of the specimen's surface. (APP.)

Schwann Cells Cells which wrap themselves around the axons of neurons forming an insulating myelin sheath composed of many layers of plasma membrane. (23)

Sclereids Irregularly-shaped sclerenchyma cells, all having thick cell walls; a component of seed coats and nuts. (18)

Sclerenchyma Component of the ground tissue system of plants. They are thick walled cells of various shapes and sizes, providing support or protection. They continue to function after the cell dies. (18)

Sclerenchyma Fibers Non-living elongated plant cells with tapering ends and thick secondary walls. A supportive cell type found in various plant tissues. (18)

Sebaceous Glands Exocrine glands of the skin that produce a mixture of lipids (sebum) that oil the hair and skin. (26)

Secondary Cell Wall An additional cell wall that improves the strength and resiliency of specialized plant cells, particularly those cells found in stems that support leaves, flowers, and fruit. (5)

Secondary Consumer Organism that feeds exclusively on primary consumers; mostly animals, but some plants. (41)

Secondary Growth Growth from cambia in perennials; results in an increase in the diameter of stems and roots. (18)

Secondary Meristems (vascular cambium, cork cambrium) Rings or clusters of meristematic cells that increase the width of stems and roots when the divide. (18)

Secondary Sex Characteristics Those characteristics other than the gonads and reproductive tract that develop in response to sex hormones. For example, breasts and pubic hair in women and a deep voice and pubic hair in men. (31)

Secondary Succession The development of a community in an area previously occupied by a community, but which was disturbed in some manner; for example, fire, development, or clear-cutting forests. (41)

Secondary Tissues Tissues produced to accommodate new cell production in plants with woody growth. Secondary tissues are produced from cambia, which produce vascular and cork tissues, leading to an increase in plant girth. (18)

Second Messenger Many hormones, such as glucagon and thyroid hormone, evoke a response by binding to the outer surface of a target cell and causing the release of another substance (which is the second messenger). The best-studied second messenger is cyclic AMP which is formed by an enzyme on the inner surface of the plasma membrane following the binding of a hormone to the outer surface of the membrane. The cyclic AMP diffuses into the cell and activates a protein kinase. (25)

Secretion The process of exporting materials produced by the cell. (5)

Seed A mature ovule consisting of the embryo, endosperm, and seed coat. (20)

Seed Dormancy Metabolic inactivity of seeds until favorable conditions promote seed germination. (20)

Secretin Hormone secreted by endocrine cells in the wall of the intestine that stimulates the release of digestive products from the pancreas. (25)

Segmentation A condition in which the body is constructed, at least in part, from a series of repeating parts. Segmentation occurs in annelids, arthropods, and vertebrates (as revealed during embryonic development). (39)

Selectively Permeable A term applied to the plasma membrane because membrane proteins control which molecules are transported. Enables a cell to import and accumulate the molecules essential for normal metabolism. (7)

Semen The fluid discharged during a male orgasm. (31)

Semiconsevative Replication The manner in which DNA replicates; half of the original DNA strand is conserved in each new double helix. (14)

Seminal Vesicles The organs which produce most of the ejaculatory fluid. (31)

Seminiferous Tubules Within the testes, highly coiled and compacted tubules, lined with a self-perpetuating layer of spermatogonia, which develop into sperm. (31)

Senescence Aging and eventual death of an organism, organ or tissue. (3, 18)

Sense Strand The one strand of a DNA double helix that contains the information that encodes the amino sequence of a polypeptide. This is the strand that is selectively transcribed by RNA polymerase forming an mRNA that can be properly translated. (14)

Sensory Neurons Neurons which relay impulses to the central nervous system. (23)

Sensory Receptors Structures that detect changes in the external and internal environment and transmit the information to the nervous system. (24)

Sepal The outermost whorl of a flower, enclosing the other flower parts as a flower bud; collectively called the calyx. (20)

Sessile Sedentary, incapable of independent movement. (39)

Sex Chromosomes The one chromosomal pair that is not identical in the karyotypes of males and females of the same animal species. (10, 13)

Sex Hormones Steroid hormones which influence the production of gametes and the development of male or female sex characteristics. (25)

Sexual Dimorphism Differences in the appearance of males and females in the same species. (33)

Sexual Reproduction The process by which haploid gametes are formed and fuse during fertilization to form a zygote. (31)

Sexual Selection The natural selection of adaptations that improve the chances for mating and reproducing. (33)

Shivering Involuntary muscular contraction for generating metabolic heat that raises body temperature. (28)

Shoot In angiosperms, the system consisting of stems, leaves, flowers and fruits. (18)

Shoot System The above-ground portion of an angiosperm plant consisting of stems with nodes, including branches, leaves, flowers and fruits. (18)

Short-Day Plants Plants that flower in late summer or fall when the length of daylight becomes shorter than some critical period. (21)

Shrubland A biome characterized by densely growing woody shrubs in mediterranean type climate; growth is so dense that understory plants are not usually present. (40)

Sickle Cell Anemia A genetic (recessive autosomal) disorder in which the beta globin genes of adult hemoglobin molecules contain an amino acid substitution which alters the ability of hemoglobin to transport oxygen. During times of oxygen stress, the red blood cells of these individuals may become sickle shaped, which interferes with the flow of the cells through small blood vessels. (4, 17)

Sieve Plate Found in phloem tissue in plants, the wall between sieve-tube members, containing perforated areas for passage of materials. (18)

Sieve-Tube Member A living, food-conducting cell found in phloem tissue of plants; associated with a companion cell. (18)

Sigmoid Growth Curve An S-shaped curve illustrating the lag phase, exponential growth, and eventual approach of a population to its carrying capacity. (43)

Sign Stimulus An object or action in the environment that triggers an innate behavior. (44)

Simple Fruits Fruits that develop from the ovary of one pistil. (20)

Simple Leaf A leaf that is undivided; only one blade attached to the petiole. (18)

Sinoatrial (SA) Node A collection of cells that generates an action potential regulating heart beat; the heart's pacemaker. (28)

Skeletal Muscles Separate bundles of parallel, striated muscle fibers anchored to the bone, which they can move in a coordinated fashion. They are under voluntary control. (26)

Skeleton A rigid form of support found in most animals either surrounding the body with a protective encasement or providing a living girder system within the animal. (26)

Skull The bones of the head, including the cranium. (26)

Slow-Twitch Fibers Skeletal muscle fibers that depend on aerobic metabolism for ATP production. These fibers are capable of undergoing contraction for extended periods of time without fatigue, but generate lesser forces than fast-twitch fibers. (9)

Small Intestine Portion of the intestine in which most of the digestion and absorption of nutrients takes place. It is so named because of its narrow diameter. There are three sections: duodenum, jejunum, and ilium. (27)

Smell Sense of the chemical composition of the environment. (24)

Smooth ER (SER) Membranes of the endoplasmic reticulum that have no ribosomes on their surface. SER is generally more tubular than the RER. Often acts to store calcium or synthesize steroids. (5)

Smooth Muscle The muscles of the internal organs (digestive tract, glands, etc.). Composed of spindle-shaped cells that interlace to form sheets of visceral muscle. (26)

Social Behavior Behavior among animals that live in groups composed of individuals that are dependent on one another and with whom they have evolved mechanisms of communication. (44)

Social Learning Learning of a behavior from other members of the species. (44)

Social Parasitism Parasites that use behavioral mechanisms of the host organism to the parasite's advantage, thereby harming the host. (42)

Solute A substance dissolved in a solvent. (3)

Solution The resulting mixture of a solvent and a solute. (3)

Solvent A substance in which another material dissolves by separating into individual molecules or ions. (3)

Somatic Cells Cells that do not have the potential to form reproductive cells (gametes). Includes all cells of the body except germ cells. (11)

Somatic Nervous System The nerves that carry messages to the muscles that move the skeleton either voluntarily or by reflex. (23)

Somatic Sensory Receptors Receptors that respond to chemicals, pressure, and temperature that are present in the skin, muscles, tendons, and joints. Provides a sense of the physiological state of the body. (24)

Somites In the vertebrate embryo, blocks of mesoderm on either side of the notochord that give rise to muscles, bones, and dermis. (32)

Speciation The formation of new species. Occurs when one population splits into separate populations that diverge genetically to the point where they become separate species. (33)

Species Taxonomic subdivisions of a genus. Each species has recognizable features that distinguish it from every other species. Members of one species generally will not interbreed with members of other species. (33)

Specific Epithet In taxonomy, the second term in an organism's scientific name identifying its species within a particular genus. (1)

Spermatid Male germ cell that has completed meiosis but has not yet differentiated into a sperm. (31)

Spermatogenesis The production of sperm. (31)

Spermatogonia Male germ cells that have not yet begun meiosis. (31)

Spermatozoa (Sperm) Male gametes. (31)

Sphinctors Circularly arranged muscles that close off the various tubes in the body.

Spinal Cord A centralized mass of neurons for processing neurological messages and linking the brain with that part of peripheral nervous system not reached by the cranial nerves. (23)

Spinal Nerves Paired nerves which emerge from the spinal cord and innervate the body. Humans have 31 pairs of spinal nerves. (23)

Spindle Apparatus In dividing eukaryotic cells, the complex rigging, made of microtubules, that aligns and separates duplicated chromosomes. (10)

Splash Zone In the intertidal zone, the uppermost region receiving splashes and sprays of water to the mean of high tides. (40)

Spleen One of the organs of the lymphatic system that produces lymphocytes and filters blood; also produces red blood cells in the human fetus. (28)

Splicing The step during RNA processing in which the coding segments of the primary transcript are covalently linked together to form the mRNA. (15)

Spongy Parenchyma In monocot and dicot leaves, loosely arranged cells functioning in photosynthesis. Found above the lower epidermis and beneath the palisade parenchyma in dicots, and between the upper and lower epidermis in monocots. (18)

Spontaneous Generation Disproven concept that living organisms can arise directly from inanimate materials. (2)

Sporangiospores Black, asexual spores of the zygomycete fungi. (37)

Sporangium A hollow structure in which spores are formed. (37)

Spores In plants, haploid cells that develop into the gametophyte generation. In fungi, an asexual or sexual reproductive cell that gives rise to a new mycelium. Spores are often lightweight for their dispersal and adapted for survival in adverse conditions. (37)

Sporophyte The diploid spore producing generation in plants. (38)

Stabilizing Selection Natural selection favoring an intermediate phenotype over the extremes. (33)

Starch Polysaccharides used by plants to store energy. (4)

Stamen The flower's male reproductive organ, consisting of the pollen-producing anther supported by a slender stalk, the filament. (20)

Stem In plants, the organ that supports the leaves, flowers, and fruits. (18)

Stem Cells Cells which are undifferentiated and capable of giving rise to a variety of different types of differentiated cells. For example, hematopoietic stem cells are capable of giving rise to both red and white blood cells. (17)

Steroids Compounds classified as lipids which have the basic four-ringed molecular skeleton as represented by cholesterol. Two examples of steroid hormones are the sex hormones; testosterone in males and estrogen in females. (4, 25)

Stigma The sticky area at the top of each pistil to which pollen adheres. (20)

Stimulus Any change in the internal or external environment to which an organism can respond. (24)

Stomach A muscular sac that is part of the digestive system where food received from the esophagus is stored and mixed, some breakdown of food occurs, and the chemical degradation of nutrients begins. (27)

Stomates (Pl. Stomata) Microscopic pores in the epidermis of the leaves and stems which allow gases to be exchanged between the plant and the external environment. (18)

Stratified Epithelia Multicellular layered epithelium. (22)

Stream Flowing body of surface fresh water; streams merge together into larger streams and rivers. (40)

Stretch Receptors Sensory receptors embedded in muscle tissue enabling muscles to respond reflexively when stretched. (23, 24)

Striated Referring to the striped appearance of skeletal and cardiac muscle fibers. (26)

Strobilus In lycopids, terminal, cone-like clusters of specialized leaves that produce sporangia.

Stroma The fluid interior of chloroplasts. (8)

Stromatolites Rocks formed from masses of dense bacteria and mineral deposits. Some of these rocky masses contain cells that date back over three billion years revealing the nature of early prokaryotic life forms. (35)

Structural Genes DNA segments in bacteria that direct the formation of enzymes or structural proteins. (15)

Style The portion of a pistil which joins the stigma to the ovary. (20)

Substrate-Level Phosphorylation The formation of ATP by direct transfer of a phosphate group from a substrate, such as a sugar phosphate, to ADP. ATP is formed without the involvement of an electron transport system. (9)

Substrates The reactants which bind to enzymes and are subsequently converted to products. (6)

Succession The orderly progression of communities leading to a climax community. It is one of two types: primary, which occurs in areas where no community existed before; and secondary, which occurs in disturbed habitats where some soil and perhaps some organisms remain after the disturbance. (41)

Succulents Plants having fleshy, water-storing stems or leaves. (40)

Suppressor T Cells A class of T cells that regulate immune responses by inhibiting the activation of other lymphocytes. (30)

Surface Area-to-Volume Ratio The ratio of the surface area of an organism to its volume, which determines the rate of exchange of materials between the organism and its environment. (22)

Surface Tension The resistance of a liquid's surface to being disrupted. In aqueous solutions, it is caused by the attraction between water molecules. (3)

Survivorship Curve Graph of life expectancy, plotted as the number of survivors versus age. (43)

Sweat Glands Exocrine glands of the skin that produce a dilute salt solution, whose evaporation cools in the body. (26)

Symbiosis A close, long-term relationship between two individuals of different species. (42)

Symmetry Referring to a body form that can be divided into mirror image halves by at least one plane through its body. (39)

Sympathetic Nervous System Part of the autonomic nervous system that tends to stimulate bodily activities, particularly those involved with coping with stressful situations. (23)

Sympatric Speciation Speciation that occurs in populations with overlapping distributions. It is common in plants when polyploidy arises within a population. (33)

Synapse Juncture of a neuron and its target cell (another neuron, muscle fiber, gland cell). (23)

Synapsis The pairing of homologous chromosomes during prophase of meiosis I. (11)

Synaptic Cleft Small space between the synaptic knobs of a neuron and its target cell. (23)

Synaptic Knobs The swellings that branch from the end of the axon. They deliver the neurological impulse to the target cell. (23)

Synaptonemal Complex Ladderlike structure that holds homologous chromosomes together as a tetrad during crossing over in prophase I of meiosis. (11)

Synovial Cavities Fluid-filled sacs around joints, the function of which is to lubricate and separate articulating bone surfaces. (26)

Systemic Circulation Part of the circulatory system that delivers oxygenated blood to the tissues and routes deoxygenated blood back to the heart. (28)

Systolic Pressure The first number of a blood pressure reading; the highest pressure attained in the arteries as blood is propelled out of the heart. (28)

◄ T ►

Taiga A biome found south of tundra biomes; characterized by coniferous forests, abundant precipitation, and soils that thaw only in the summer. (40)

Tap Root System Root system of plants having one main root and many smaller lateral roots. Typical of conifers and dicots. (18)

Taste Sense of the chemical composition of food. (24)

Taxonomy The science of classifying and grouping organisms based on their morphology and evolution. (1)

T Cell Lymphocytes that carry out cell-mediated immunity. They respond to antigen stimulation by becoming helper cells, killer cells, and memory cells. (30)

Telophase The final stage of mitosis which begins when the chromosomes reach their spindle poles and ends when cytokinesis is completed and two daughter cells are produced. (10)

Tendon A dense connective tissue cord that connects a skeletal muscle to a bone. (26)

Teratogenic Embryo deforming. Chemicals such as thalidomide or alcohol are teratogenic because they disturb embryonic development and lead to the formation of an abnormal embryo and fetus. (32)

Terminal Electron Acceptor In aerobic respiration, the molecule of O_2 which removes the electron pair from the final cytochrome of the respiratory chain. (9)

Terrestrial Living on land. (40)

Territory (Territoriality) An area that an animal defends against intruders, generally in the protection of some resource. (42, 44)

Tertiary Consumer Animals that feed on secondary consumers (plant or animal) or animals only. (41)

Test Cross An experimental procedure in which an individual exhibiting a dominant trait is crossed to a homozygous recessive to determine whether the first individual is homozygous or heterozygous. (12)

Testis In animals, the sperm-producing gonad of the male. (23)

Testosterone The male sex hormone secreted by the testes when stimulated by pituitary gonadotropins. (31)

Tetrad A unit of four chromatids formed by a synapsed pair of homologous chromosomes, each of which has two chromatids. (11)

Thallus In liverworts, the flat, ground-hugging plant body that lacks roots, stems, leaves, and vascular tissues. (38)

Theory of Tolerance Distribution, abundance and existence of species in an ecosystem are determined by the species' range of tolerance of chemical and physical factors. (41)

Thermoreceptors Sensory receptors that respond to changes in temperature. (24)

Thermoregulation The process of maintaining a constant internal body temperature in spite of fluctuations in external temperatures. (28)

Thigmotropism Changes in plant growth stimulated by contact with another object, e.g., vines climbing on cement walls. (21)

Thoracic Cavity The anterior portion of the body cavity in which the lungs are suspended. (39)

Thylakoids Flattened membrane sacs within the chloroplast. Embedded in these membranes are the light-capturing pigments and other components that carry out the light-dependent reactions of photosynthesis. (8)

Thymus Endocrine gland in the chest where T cells mature. (30)

Thyroid Gland A butterfly-shaped gland that lies just in front of the human windpipe, producing two metabolism-regulating hormones, thyroxin and triodothyronine. (25)

Thyroid Hormone A mixture of two iodinated amino acid hormones (thyroxin and triiodothyronine) secreted by the thyroid gland. (25)

Thyroid Stimulating Hormone (TSH) An anterior pituitary hormone which stimulates secretion by the thyroid gland. (25)

Tissue An organized group of cells with a similar structure and a common function. (22)

Tissue System Continuous tissues organized to perform a specific function in plants. The three plant tissue systems are: dermal, vascular, and ground (fundamental). (18)

Tolerance Range The range between the maximum and minimum limits for an environmental factor that is necessary for an organism's survival. (41)

Totipotent The genetic potential for one type of cell from a multicellular organism to give rise to any of the organism's cell types, even to generate a whole new organism. (15)

Trachea The windpipe; a portion of the respiratory tract between the larynx and bronchii. (29)

Tracheal Respiratory System A network of tubes (tracheae) and tubules (tracheoles) that carry air from the outside environment directly to the cells of the body without involving the circulatory system. (29, 39)

Tracheid A type of conducting cell found in xylem functioning when a cell is dead to transport water and dissolved minerals through its hollow interior. (18)

Tracheophytes Vascular plants that contain fluid-conducting vessels. (38)

Transcription The process by which a strand of RNA assembles along one of the DNA strands. (14)

Transcriptional-Level Control Control of gene expression by regulating whether or not a specific gene is transcribed and how often. (15)

Transduction A type of genetic recombination resulting from transfer of genes from one organism to another by a virus.

Transfer RNA (tRNA) A type of RNA that decodes mRNA's codon message and translates it into amino acids. (14)

Transgenic Organism An organism that possesses genes derived from a different species. For example, a sheep that carries a human gene and secretes the human protein in its milk is a transgenic animal. (16)

Translation The cell process that converts a sequence of nucleotides in mRNA into a sequence of amino acids in a polypeptide. (14)

Translational-Level Control Control of gene expression by regulating whether or not a specific mRNA is translated into a polypeptide. (15)

Translocation The joining of segments of two nonhomologous chromosomes (13)

Transmission Electron Microscope (TEM) A microscope that works by shooting electrons through very thinly sliced specimens. The result is an enormously magnified image, two-dimensional, of remarkable detail. (App.)

Transpiration Water vapor loss from plant surfaces. (19)

Transpiration Pull The principle means of water and mineral transport in plants, initiated by transpiration. (19)

Transposition The phenomenon in which certain DNA segments (mobile genetic elements, or jumping genes) tend to move from one part of the genome to another part. (15)

Transverse Fission The division pattern in ciliated protozoans where the plane of division is perpendicular to the cell's length.

Trimester Each of the three stages comprising the 266-day period between conception and birth in humans. (32)

Triploid Having three sets of chromosomes, abbreviated 3N. (11)

Trisomy Three copies of a particular chromosome per cell. (17)

Trophic Level Each step along a feeding pathway. (41)

Trophozoite The actively growing stage of polymorphic protozoa. (37)

Tropical Rain Forest Lush forests that occur near the equator; characterized by high annual rainfall and high average temperature. (40)

Tropical Thornwood A type of shrubland occurring in tropical regions with a short rainy season. Plants lose their small leaves during dry seasons, leaving sharp thorns. (40)

Tropic Hormones Hormones that act on endocrine glands to stimulate the production and release of other hormones. (25)

Tropisms Changes in the direction of plant growth in response to environmental stimuli, e.g., light, gravity, touch. (21)

True-Breeder Organisms that, when bred with themselves, always produce offspring identical to the parent for a given trait. (12)

Tubular Reabsorption The process by which substances are selectively returned from the fluid in the nephron to the bloodstream. (28)

Tubular Secretion The process by which substances are actively and selectively transported from the blood into the fluid of the nephron. (28)

Tumor-Infiltrating Lymphocytes (TILs) Cytotoxic T cells found within a tumor mass that have the capability to specifically destroy the tumor cells. (30)

Tumor-Suppressor Genes Genes whose products act to block the formation of cancers. Cancers form only when both copies of these genes (one on each homologue) are mutated. (13)

Tundra The marshy, unforested biome in the arctic and at high elevations. Frigid temperatures for most of the year prevent the subsoil from thawing, which produces marshes and ponds. Dominant vegetation includes low growing plants, lichens, and mosses. (40)

Turgor Pressure The internal pressure in a plant cell caused by the diffusion of water into the cell. Because of the rigid cell wall, pressure can increase to where it eventually stops the influx of more water. (7)

Turner Syndrome A person whose cells have only one X chromosome and no second sex chromosome (XO). These individuals develop as immature females. (17)

◄ U ►

Ultimate (Top) Consumer The final carnivore trophic level organism, or organisms that escaped predation; these consumers die and are eventually consumed by decomposers. (41)

Ultracentrifuge An instrument capable of spinning tubes at very high speeds, delivering centrifugal forces over 100,000 times the force of gravity. (9)

Unicellular The description of an organism where the cell is the organism. (35)

Uniform Pattern Distribution of individuals of a population in a uniform arrangement, such as individual plants of one species uniformly spaced across a region. (43)

Urethra In mammals, a tube that extends from the urinary bladder to the outside. (28)

Urinary Tract The structures that form and export urine: kidneys, ureters, urinary bladder, and urethra. (28)

Urine The excretory fluid consisting of urea, other nitrogenous substances, and salts dissolved in water. It is formed by the kidneys. (28)

Uterine (Menstrual) Cycle The repetitive monthly changes in the uterus that prepare the endometrium for receiving and sustaining an embryo. (31)

Uterus An organ in the female reproductive system in which an embryo implants and is maintained during development. (31)

◄ V ►

Vaccines Modified forms of disease-causing microbes which cannot cause disease but retain the same antigens of it. They permit the immune system to build memory cells without diseases developing during the primary immune response. (30)

Vacoconstriction Reduction in the diameter of blood vessels, particularly arterioles. (28)

Vacuole A large organelle found in mature plant cells, occupying most of the cell's volume, sometimes more than 90% of it. (5)

Vagina The female mammal's copulatory organ and birth canal. (31)

Variable (Experimental) A factor in an experiment that is subject to change, i.e., can occur in more than one state. (2)

Vascular Bundles Groups of vascular tissues (xylem and phloem) in the shoot of a plant. (19)

Vascular Cambrium In perennials, a secondary meristem that produces new vascular tissues. (18)

Vascular Cylinder Groups of vascular tissues in the central region of the root. (18)

Vascular Plants Plants having a specialized conducting system of vessels and tubes for transporting water, minerals, food, etc., from one region to another. (18)

Vascular Tissue System All the vascular tissues in a plant, including xylem, phloem, and the vascular cambium or procambium. (18)

Vasodilation Increase in the diameter of blood vessels, particularly arterioles. (28)

Veins In plants, vascular bundles in leaves. In animals, blood vessels that return blood to the heart. (28)

Venation The pattern of vein arrangement in leaf blades. (18)

Ventricle Lower chamber of the heart which pumps blood through arteries. There is one ventricle in the heart of lower vertebrates and two ventricles in the four-chambered heart of birds and mammals. (28)

Venules Small veins that collect blood from the capillaries. They empty into larger veins for return to the heart. (28)

Vertebrae The bones that form the backbone. In the human there are 33 bones arranged in a gracefully curved line along the bone, cushioned from one another by disks of cartilage. (26)

Vertebral Column The backbone, which encases and protects the spinal cord. (26)

Vertebrates Animals with a backbone. (39)

Vesicles Small membrane-enclosed sacs which form from the ER and Golgi complex. Some store chemicals in the cells; others move to the surface and fuse with the plasma membrane to secrete their contents to the outside. (5)

Vessel A tube or connecting duct containing or circulating body fluids. (18)

Vessel Member A type of conducting cell in xylem functioning when the cell is dead to transport water and dissolved minerals through its hollow interior; also called a vessel element. (18)

Vestibular Apparatus A portion of the inner ear of vertebrates that gathers information about the position and movement of the head for use in maintaining balance and equilibrium. (24)

Vestigial Structure Remains of ancestral structures or organs which were, at one time, useful. (34)

Villi Finger-like projections of the intestinal wall that increase the absorption surface of the intestine. (27)

Viroids are associated with certain diseases of plants. Each viroid consists solely of a small single-stranded circle of RNA unprotected by a protein coat. (36)

Virus Minute structures composed of only heredity information (DNA or RNA), surrounded by a protein or protein/lipid coat. After infection, the viral mucleic acid subverts the metabolism of the host cell, which then manufactures new virus particles. (36)

Visible Light The portion of the electromagnetic spectrum producing radiation from 380 nm to 750 nm detectable by the human eye.

Vitamins Any of a group of organic compounds essential in small quantities for normal metabolism. (27)

Vocal Cords Muscular folds located in the larynx that are responsible for sound production in mammals. (29)

Vulva The collective name for the external features of the human female's genitals. (31)

◀ W ▶

Water Vascular System A system for locomotion, respiration, etc., unique to echinoderms. (39)

Wavelength The distance separating successive crests of a wave. (8)

Waxes A waterproof material composed of a number of fatty acids linked to a long chain alcohol. (4)

White Matter Regions of the brain and spinal cord containing myelinated axons, which confer the white color. (23)

Wild Type The phenotype of the typical member of a species in the wild. The standard to which mutant phenotypes are compared. (13)

Wilting Drooping of stems or leaves of a plant caused by water loss. (7)

Wood Secondary xylem. (18)

◀ X ▶

X Chromosome The sex chromosome present in two doses in cells of a female, and in one dose in the cells of a male. (13)

X-Linked Traits Traits controlled by genes located on the X chromosome. These traits are much more common in males than females. (13)

Xylem The vascular tissue that transports water and minerals from the roots to the rest of the plant. Composed of tracheids and vessel members. (18)

◀ Y ▶

Y Chromosome The sex chromosome found in the cells of a male. When Y-carrying sperm unite with an egg, all of which carry a single X chromosome, a male is produced. (13)

Y-Linked Inheritance Genes carried only on Y chromosomes. There are relatively few Y-linked traits; maleness being the most important such trait in mammals. (13)

Yeast Unicellular fungus that forms colonies similar to those of bacteria. (37)

Yolk A deposit of lipids, proteins, and other nutrients that nourishes a developing embryo. (32)

Yolk Sac A sac formed by an extraembryonic membrane. In humans, it manufactures blood cells for the early embryo and later helps to form the umbilical cord. (32)

◀ Z ▶

Zero Population Growth In a population, the result when the combined positive growth factors (births and immigration). (43)

Zooplankton Protozoa, small crustaceans and other tiny animals that drift with ocean currents and feed on phytoplankton. (37, 40)

Zygospore The diploid spores of the zygomycete fungi, which include Rhizopus, a common bread mold. After a period of dormancy, the zygospore undergoes meiosis and germinates. (36)

Zygote A fertilized egg. The diploid cell that results from the union of a sperm and egg. (32)

Photo Credits

Part 1 Opener Norbert Wu. **Chapter 1** Fig. 1.1: Jeff Gnass. Fig. 1.2a: Frans Lanting/ Minden Pictures, Inc. Fig. 1.2b: David Muench. Fig. 1.3a: Courtesy Dr. Alan Cheetham, National Museum of National History, Smithsonian Institution. Fig. 1.3b: Manfred Kage/Peter Arnold. Fig. 1.3c: Stephen Dalton/NHPA. Fig. 1.3d: Charles Summers, Jr. Fig. 1.3e: Larry West/ Photo Researchers. Fig. 1.3f: Bianca Lavies. Fig. 1.3g: Steve Allen/Peter Arnold. Fig. 1.3h: George Grall. Fig. 1.3i: Michio Hoshino/Minden Pictures, Inc. Fig. 1.6a: CNRI/Science Photo Library/Photo Researchers. Fig. 1.6b: M. Abbey/ Visuals Unlimited. Fig. 1.6c: Steve Kaufman/ Peter Arnold. Fig. 1.6d: Willard Clay. Fig. 1.6e: Jim Bradenburg/Minden Pictures, Inc. Fig. 1.7: (pages 16-17) Charles A. Mauzy; (page 16, top) Anthony Mercieca/Natural Selection; (page 16, center) Wolfgang Bayer/Bruce Coleman, Inc.; (page 16, bottom) Dr. Jeremy Burgess/Science Photo Library/Photo Researchers; (page 17, top) Dr. Eckart Pott/Bruce Coleman, Ltd.; (page 17, bottom) Art Wolfe. Fig. 1..8: Paul Chesley/ Photographers Aspen. Fig. 1.9a: Rainbird/Robert Harding Picture Library. Fig. 1.11: Dick Luria/ FPG International. **Chapter 2** Fig. 2.2a: Couresy Institut Pasteur. Fig. 2.3a: Topham/The Image Works. Fig. 2.3b: Leonard Lessin/Peter Arnold. Fig. 2.3c: Bettmann Archive. Fig. 2.5a: Laurence Gould/Earth Scenes/Animals Animals. Fig. 2.5b: Ted Horowitz/The Stock Market. **Part 2 Opener:** Nancy Kedersha. **Chapter 3** Fig. 3.1: Franklin Viola. Fig. 3.2: Courtesy Pachyderm Scientific Industries. Bioline: Jacan-Yves Kerban/The Image Bank. Fig. 3.7: Courtesy Stephen Harrison, Harvard Biochemistry Department. Fig. 3.11: Gary Milburn/Tom Stack & Associates. Fig. 3.12: Courtesy R. S. Wilcox, Biology Department, SUNY Binghampton. **Chapter 4** Fig. 4.1: Alastair Black/Tony Stone World Wide. Fig. 4.5a: Don Fawcett/Visuals Unlimited. Fig. 4.5b: Jeremy Burgess/Photo Researchers. Fig. 4.5c: Cabisco/Visuals Unlimited. Fig. 4.6: Tony Stone World Wide. Fig. 4.7: Zig Leszcynski/Animals Animals. Fig. 4.8: Robert & Linda Mitchell. Human Perspective: (a) Photofest (b) Bill Davila/Retna. Fig. 4.12a: Frans Lanting/AllStock, Inc. Fig. 4.12b: Mantis Wildlife Films/Oxford Scientific Films/Animals Animals. Fig. 4.17: Stanley Flegler/Visuals Unlimited. Fig. 4.18a: Courtesy Nelson Max, Lawrence Livermore Laboratory. Fig. 4.18b: Tsuned Hayashida/The Stock Market. **Chapter 5** Fig. 5.1a: The Granger Collection. Fig. 5.1b: Bettmann Archive. Fig. 5.2a: Dr. Jeremey Burgess/ Photo Researchers. Fig. 5.2b: CNRI/Photo Researchers. Fig. 5.4: Omikron/Photo Researchers. Fig. 5.6a: Courtesy Richard Chao, California State University at Northridge. Fig. 5.6c: Courtesy Daniel Branton, University of Berkeley. Fig. 5.7: Courtesy G. F. Bohr. Fig. 5.8: Courtesy Michael Mercer, Zoology Department, Arizona State University. Fig. 5.9: Courtesy D. W. Fawcett, Harvard Medical School. Fig. 5.10a: Courtesy U.S. Department of Agriculture, Fig. 5.11: Courtesy Dr. Birgit H. Satir, Albert Einstein College of Medicine. Fig. 5.13: Courtesy Lennart Nilsson, from *A Child Is Born*. Fig. 5.14a: K. R. Porter/Photo Researchers. Fig. 5.14c: Courtesy Lennart Nilsson, BonnierAlba. Fig. 5.14d: Courtesy Lennart Nilsson, From *A Child Is Born*. Fig. 5.15a: Courtesy J. Elliot Weier. Fig. 5.16a: Courtesy J.V. Small. Fig. 5.17a: Peter Parks/Animals Animals. Fig. 5.17b: Courtesy Dr. Manfred Hauser, RUHR-Universitat Rochum. Fig. 5.18: Courtesy Jean Paul Revel, Division of Biology, California Institute of Technology. Fig. 5.19a: Courtesy E. Vivier, from *Paramecium* by W.J. Wagtendonk, Elsevier North Holland Biomedical Press, 1974. Fig. 5.19b: David Phillips/Photo Researchers. Fig. 5.20: Courtesy C.J. Brokaw and T.F. Simonick, *Journal of Cell Biology*, 75:650 (1977). Reproduced with permission. Fig. 5.21: Courtesy L.G. Tilney and K. Fujiwara. Fig. 5.22a: Courtesy W. Gordon Whaley, University of Texas, Austin. Fig. 5.22c: Courtesy R.D. Preston, University of Leeds, London. **Chapter 6** Fig. 6.1a: Alex Kerstitch. Fig. 6.1b: Peter Parks/Earth Scenes. Fig. 6.2: Marty Stouffer/Animals Animals. Fig. 6.3: Kay Chernush/The Image Bank. Fig. 6.4b: Courtesy Computer Graphics Laboratory, University of California, San Francisco. Fig. 6.8: Courtesy Stan Koszelek, Ph.D., University of California, Riverside. Fig. 6.10: Swarthout/The Stock Market. **Chapter 7** Fig. 7.5: Ed Reschke. Fig. 7.10a: Lennart Nilsson, ©Boehringer Ingelheim, International Gmbh; from *The Incredible Machine*. Human Perspective: (Fig. 1a) Martin Rotker/Phototake; (Fig. 1b) Cabisco/Visuals Unlimited. Fig. 7.11a: Courtesy Dr. Ravi Pathak, Southwestern Medical Center, University of Texas. **Chapter 8** Fig. 8.1a: Carr Clifton. Fig. 8.1b: Shizuo Lijima/Tony Stone World Wide. Fig. 8.5: Joe Englander/ Viesti Associates, Inc. Fig. 8.6: Courtesy T. Elliot Weier. Fig. 8.10: Courtesy Lawrence Berkeley Laboratory, University of California. Fig. 8.13: Pete Winkel/Atlanta/Stock South. Bioline: Courtesy Woods Hole Oceanographic Institution. **Chapter 9** Fig. 9.1: Stephen Frink/ AllStock, Inc. Bioline: (Fig. 1) Frank Oberle/ Bruce Coleman, Inc.; (Fig. 2) Movie Stills Archive. Fig. 9.6: Topham/The Image Works. Fig. 9.9: Courtesy Gerald Karp. Human Perspective: (top and bottom insets) Courtesy MacDougal; (top) Jerry Cooke; (bottom) Richard Kane/Sportchrome East/West. Fig. 9.11: Peter Parks/NHPA. Fig. 9.12a: Grafton M. Smith/ The Image Bank. **Chapter 10** Fig. 10.1: Sipa Press. Fig. 10.2a: Institut Pasteur/CNRI/ Phototake. Fig. 10.3a: Dr. R. Vernon/Phototake. Fig. 10.3b: CNRI/Science Photo Library/Photo Researchers. Fig. 10.6: Courtesy Dr. Andrew Bajer, University of Oregon, Fig. 10.7: CNRI/ Science Photo Library/Photo Researchers. Fig. 10.8: Dr. G. Shatten/Science Photo Library/ Photo Researchers. Fig. 10.9: Courtesy Professor R.G. Kessel. Fig. 10.10: David Phillips/ Photo Researchers. **Chapter 11** Fig. 11.3: Courtesy Science Software Systems. Fig. 11.5a: Courtesy Dr. A.J. Solari, from *Chromosoma*, vol. 81, p. 330 (1980), Human Perspective: Donna Zweig/Retna. Fig. 11.7: Cabisco/Visuals Unlimited. **Part 3 Opener:** David Scharf. **Chapter 12** Fig. 12.1: Courtesy Dr. Ing Jaroslav Krizenecky. Fig. 12.6a: Doc Pele/Retna. Fig. 12.6b: FPG International. Fig. 12.9: Sydney Freelance/Gamma Liaison. Fig. 12.11 Hans Reinhard/Bruce Coleman, Inc. **Chapter 13** Fig. 13.2a: Robert Noonan. Fig. 13.6: Biological Photo Service. Bioline: Norbert Wu. Fig. 13.9: Historical Pictures Service. Fig. 13.11a: Courtesy M.L. Barr. Fig. 13.11b: Jean Pragen/Tony Stone World Wide. Fig. 13.12: Courtesy Lawrence Livermore National Laboratory. **Chapter 14** Fig. 14.1b: Lee D. Simon/Science Photo Library/Photo Researchers. Fig. 14.4: David Leah/Science Photo Library/Photo Researchers. Fig. 14.5: Dr. Gopal Murti/Science Photo Library/Photo Researchers. Fig. 14.6b: Fawcett/ Olins/Photo Researchers. Fig. 14.7: Courtesy U.K. Laemmli. Fig. 14.10a: From M. Schnos and.R.B. Inman, *Journal of Molecular Biology*, 51:61-73 (1970), ©Academic Press. Fig. 14.10b: Courtesy Professor Joel Huberman, Roswell Park Memorial Institute. Human Perspective: (Fig. 2) Courtesy Skin Cancer Foundation; (Fig. 3) Mark Lewis/Gamma Liaison. Fig. 14.17: Courtesy Dr. O.L. Miller, Oak Ridge National Laboratory. **Chapter 15** Fig. 15.1: Courtesy Richard Goss, Brown University. Fig. 15.2a: (top left) Oxford Scientific Films/ Animals Animals; (top right) F. Stuart Westmorland/Tom Stack & Associates. Fig. 15.2b: Courtesy Dr. Cecilio Barrera, Department of Biology, New Mexico State University. Fig. 15.3b: Courtesy Michael Pique, Research Institute of Scripps Clinic. Fig. 15.7b: Courtesy Wen Su and Harrison Echols, University of California, Berkeley. Fig. 15.8a: Courtesy Stephen Case, University of Mississippi Medical Center. Fig. 15.9: Roy Morsch/The Stock Market. **Chapter 16** Fig. 16.1a: David M. Dennis/Tom Stack & Associates. Fig. 16.1b: Courtesy Lakshmi Bhatnagor, Ph.D., Michigan Biotechnology Institute. Fig. 16.2: Art Wolfe/All Stock, Inc. Fig. 16.3: Ken Graham. Fig. 16.4a: Courtesy R.L. Brinster, Laboratory for Reproductive Physiology, University of Pennsylvania. Fig. 16.4b: John Marmaras/Woodfin Camp & Associates. Fig. 16.5: Courtesy Robert Hammer, School of Veterinary Medicine, University of

Pennsylvania. Human Perspective: Courtesy Dr. James Asher, Michigan State University. Fig. 16.6b: Professor Stanley N. Cohen/Photo Researchers. Fig. 16.10: Ted Speigel/Black Star. Fig. 16.11b: Philippe Plailly/Science Photo Library/Photo Researchers. Bioline: Photograph by Anita Corbin and John O'Grady, courtesy the British Council. **Chapter 17** ig. 17.2d: Courtesy Howard Hughs Medical Institution. Fig. 17.3: Tom Raymond/Medichrome/The Stock Shop. Fig. 17.7a: Culver Pictures. Fig. 17.7b: Courtesy Roy Gumpel. Fig. 17.9a: Will & Deni McIntyre/Photo Researchers. Human Perspective: Gamma Liaison. **Part 4 Opener:** J. H. Carmichael, Jr./The Image Bank. **Chapter 18** Fig. 18.1a: Ralph Perry/Black Star. Fig. 18.1b: Jeffrey Hutcherson/DRK Photo. Fig. 18.4: Courtesy Professor Ray Evert, University of Wisconsin, Madison. Fig. 18.5: Walter H. Hodge/Peter Arnold; (inset) Courtesy Carolina Biological Supply Company. Bioline: (Fig. 1) Doug Wilson/West Light; (Fig. 2) Courtesy Gil Brum. Fig. 18.7: Luiz C. Marigo/Peter Arnold. Fig. 18.8: Dr. Jeremy Burgess/Science Photo Library/Photo Researchers. Fig. 18.9: Saul Mayer/The Stock Market; (inset) David Scharf/ Peter Arnold. Fig. 18.10a: Carr Clifton. Fig. 18.10b: Ed Reschke. Fig. 18.11: Courtesy Thomas A. Kuster, Forest Products Laboratory/ USDA. Fig. 18.12: A.J. Belling/Photo Researchers. Fig. 18.13: Courtesy DSIR Library Centre. Fig. 18.15: Biophoto Associates/Photo Researchers; (inset) Courtesy C.Y. Shih and R.G. Kessel, reproduced from *Living Images*, Science Books International, 1982. Fig. 18.16: Courtesy Industrial Research Ltd, New Zealand. Fig. 18.19: Robert & Linda Mitchell; (inset) From *Botany Principles* by Roy H. Saigo and Barbara Woodworth Saigo, ©1983 Prentice-Hall, Inc. Reproduced with permission. Fig. 18.22a: Jerome Wexler/Photo Researchers. Fig. 18.22b: Fritz Polking/Peter Arnold. Fig. 18.22c: Dwight R. Kuhn. Fig. 18.22d: Richard Kolar/Earth Scenes. Fig. 18.24: Biophoto Associates/Science Source/Photo Researchers. **Chapter 19** Fig. 19.1a: Courtesy of C.P. Reid, School of Natural Resources, University of Arizona. Fig. 19.3a: Peter Beck/The Stock Market. Fig. 19.3b: Dr. J. Burgess/Photo Researchers. Fig. 19.4: D. Cavagnaro/DRK Photo. Fig. 19.7a: Biophoto Associates/Photo Researchers. Fig. 19.7b: Alfred Pasieka/Peter Arnold. Fig. 19.7c: Courtesy DSIR Library Centre. Fig. 19.8: Martin Zimmerman. **Chapter 20** Fig. 20.1: G.I. Bernard/ Earth Scenes/Animals Animals. Fig. 20.2a: Dwight R. Kuhn/DRK Photo. Bioline (Cases 1 and 3) Biological Photo Service; (Case 2) P.H. and S.L. Ward/Natural Science Photos; (Case 4): G.I. Bernard/Animals Animals. Fig. 20.3: E.R. Degginger. Fig. 20,4a: William E. Ferguson. Fig. 20.4b: Phil Degginger. Fig. 20.4c: E.R. Degginger. Fig. 20.5a: Runk/Shoenberger/ Grant Heilman Photography. Fig. 20.5b: William E. Ferguson. Fig. 20.6a: Robert Harding Picture Library. Fig. 20,6b Richard Parker/ Photo Researchers. Fig. 20.6c: Jack Wilburn/ Earth Scenes/Animals Animals. Fig. 20.9a: Visuals Unlimted. Fig. 20.9b: Manfred Kage/ Peter Arnold. Human Perspective: Harold Sund/ The Image Bank. Fig. 20.15: Courtesy Media Resources, California State Polytechnic University. Fig. 20.16a: John Fowler/Valan Photos. Fig. 20.16b: Richard Kolar/Earth Scenes/ Animals Animals. Fig. 20.17: Breck Kent. **Chapter 21** Fig. 21.1a: Gary Milburn/Tom Stack & Associates. Fig. 21.1b: Schafer & Hill/Peter Arnold. Human Perspective: (Fig. 1) Enrico Ferorelli; (Fig. 2) Michael Nichols/ Magnum Photos, Inc. Fig. 21.5: Runk/Schoenberger/Grant Heilman Photography. Fig. 21.6: Courtesy Dr. Harlan K. Pratt, University of California, Davis. Fig. 21.7: Scott Camazine/ Photo Researchers. Fig. 21.9a: E.R. Degginger. Fig. 21.9b: Fritze Prenze/Earth Scenes/Animals Animals. Fig. 21.12a: Pam Hickman/Valan Photos. Fig. 21.12b: R.F. Head/Earth Scenes/ Animals Animals. **Part 5 Opener:** Joe Devenney/The Image Bank. **Chapter 22** Fig. 22.2b: Dr. Mary Notter/Phototake. Fig. 22.3b: Professor P. Motta, Department of Anatomy, University of "La Sapienza", Rome/Photo Researchers. Fig. 22.6a: Fred Bavendam/Peter Arnold. Fig. 22.6b: Michael Fogden/DRK Photo. Fig. 22.6c: Peter Lamberti/Tony Stone World Wide. Fig. 22.6d: Gerry Ellis Nature Photography. Fig. 22.7a: Mike Severns/Tom Stack & Associates. Fig. 22.7b: Norbert Wu. **Chapter 23** ig 23.1:Fawcett/Coggeshall/Photo Researchers. Fig. 23.5: (top left) Vu/©T. Reese and D.W. Fawcett/Visuals Unlimited; (bottom left) Courtesy Lennart Nilsson, *Behold Man*, Little Brown & Co., Boston. Fig. 23.9: Courtesy Lennart Nilsson, *Behold Man*, Little, Brown and Co., Boston. Fig. 23.10b: Alan & Sandy Carey. Human Perspective: Science Vu/Visuals Unlimited. Fig. 23.14: Focus on Sports. Fig. 23.15: The Image Works. Fig. 23.20a: Doug Wechsler/Animals Animals. Fig. 23.20b: Robert F. Sisson/National Geographic Society. **Chapter 24** Fig. 24.1a: Anthony Bannister/NHPA. Fig. 24.1b: Giddings/The Image Bank. Fig. 24.1c: Michael Fogden/Bruce Coleman, Inc. Fig. 24.4: (left) Omikron/Photo Researchers. Fig. 24.4: (inset) Don Fawcett/K. Saito/K. Hama/ Photo Researchers. Fig. 24.5a: (top) Courtesy Lennart Nilsson, *Behold Man*, Little Brown & Co., Boston; (center) Don Fawcett/K. Saito/K. Hama/Photo Researchers. Fig. 24.5b: Star File. Fig. 24.6: Philippe Petit/Sipa Press. Fig. 24.7b: Jerome Shaw. Fig. 24.10a: Oxford Scientific Films/Animals Animals. Fig. 24.10b: Kjell Sandved/Photo Researchers. Fig. 24.10c: George Shelley; (inset) Raymond A. Mendez/Animals Animals. **Chapter 25** Fig. 25.1: Robert & Linda Mitchell. Fig. 25.7: Courtesy Circus World Museum. Fig. 25.8: Courtesy A.I. Mindelhoff and D.E.. Smith, *American Journal of Medicine*, 20: 133 (1956). Fig. 25.13: Graphics by T. Hynes and A.M. de Vos using University of California at San Francisco MIDAS-plus software; photo courtesy of Genentech, Inc. **Chapter 26** Fig. 26.1: Ed Reschke/Peter Arnold. Fig. 26.2a: Joe Devenney/The Image Bank. Fig. 26.2b: John Cancalosi/DRK Photo. Fig. 26.3: Wallin/Taurus Photos. Fig. 26.4a: E.S. Ross/ Phototake. Fig. 26.5a: D. Holden Bailey/Tom Stack & Associates. Fig. 26.5b: Jany Sauvanet/Photo Researchers. Fig. 26.6a: (top) Courtesy Lennart Nilsson, *Behold Man*, Little Brown & Co., Boston; (center) Michael Abbey/ Science Source/Photo Researchers. Fig. 26.6b: F. & A. Michler/Peter Arnold. Human Perspective: Courtesy Professor Philip Osdoby, Department of Biology, Washington University, St. Louis. Fig. 26.8: Duomo Photography, Inc. Fig. 26.12: A. M. Siegelman/FPG International. Fig. 26.19a: Alese & Mort Pechter/The Stock Market. Fig. 26.19b-c: Stephen Dalton/NHPA. **Chapter 27** Bioline, Fig. 27.3, and Fig. 27.5: Courtesy Lennart Nilsson, *Behold Man*, Little Brown & Co., Boston, Fig. 27.6: Micrograph by S.L. Palay, courtesy D.W. Fawcett, from *The Cell*, © W..B. Saunders Co. Fig. 27.8a: Courtesy Gregory Antipa, San Francisco State University. Fig. 27.8b: Gregory Ochocki/Photo Researchers. Human Perspective: Derik Muray Photography, Inc. Fig. 27.9a: Warren Garst/ Tom Stack & Associates. Fig. 27.9b: Hervé Chaumeton/Jacana. Fig. 27.9c: D. Wrobel/Biological Photo Service. Fig. 27.10: Sari Levin. **Chapter 28** Fig. 28.2: Biophoto Associates/ Photo Researchers. Fig. 28.4: Courtesy Lennart Nilsson, *Behold Man*, LIttle Brown & Co., Boston. Fig. 28.10: CNRI/Science Photo Library/ Photo Researchers. Fig. 28.11a-b: Howard Sochurek. Fig. 28.11c: Dan McCoy/Rainbow. Fig. 28.12: Jean-Claude Revy/Phototake. Fig. 28.15: Dr. Tony Brain/Science Photo Library/Photo Researchers. Fig. 28.16: Alan Kearney/FPG International. Fig. 28.17a: Ed Reschke/Peter Arnold. Fig. 28.18: Manfred Kage/Peter Arnold. Fig. 28.19b: Runk/Schoenberger/Grant Heilman Photography. **Chapter 29** Fig. 29.1: Richard Kane/Sports Chrome, Inc. Fig. 29.2: Courtesy Ewald R. Weibel, Anatomisches Institut der Universität Bern, Switzerland, from "Morphological Basis of Alveolar-Capillary Gas Exchange", *Physiological Review*, Vol. 53, No. 2, April 1973, p. 425. Fig. 29.3: (top left and right insets) Courtesy Lennart Nilsson, *Behold Man*, Little Brown & Co., Boston; (center) CNRI/ Science Photo Library/Photo Researchers. Fig. 29.7: Four By Five/SUPERSTOCK. Human Perspective: (Fig. 2) Vu/O. Auerbach/Visusals Unlimited; (Fig. 3) Photofest. Fig. 29.8: Kjell Sandved/Bruce Coleman, Inc. Fig. 29.10: Robert F. Sisson/National Geographic Society. Fig. 29.11a: Robert & Linda Mitchell. **Chapter 30** Fig. 30.1 and Fig. 30.5: Courtesy Lennart Nilsson, Boehringer Ingelheim International GmbH. Bioline: Courtesy Steven Rosenberg, National Cancer Institute. Fig. 30.6: G. Robert Bishop/AllStock, Inc. Fig. 30.7b: Courtesy A.J. Olson, TSRI, Scripps Institution of Oceanography. Fig. 30.10: Alan S. Rosenthal, from "Regulation of the Immune Response—Role of the Macrophage', *New England Journal of Medicine*, November 1980, Vol. 303, #20, p. 1154, courtesy Merck Institute for Therapeutic Research. Fig. 30.11: Nancy Kedersha. Human Perspective: (Fig. 1a) David Scharf/Peter Arnold; (Fig. 1b) Courtesy Acarology Laboratory, Museum of Biological Diversity, Ohio State University, Columbus; (Fig. 1c) Scott Camazine/

Photo Researchers; (Fig. 2) Courtesy Lennart Nilsson, Boehringer Ingelheim International GmbH. **Chapter 31** Fig. 31.1a: R. La Salle/Valan Photos. Fig. 31.1b: Richard Campbell/Biological Photo Service. Fig. 31.1c: Robert Harding Picture Library. Fig. 31.2a: Doug Perrine/DRK Photo. Fig. 31.2b: George Grall. Fig. 31.2c: Biological Photo Service. Fig. 31.2d: Densey Clyne/Ocford Scientific Films/Animals Animals. Fig. 31.3a: Chuck Nicklin. Fig. 31.3b: Robert & Linda Mitchell. Bioline: Hans Pfletschinger/Peter Arnold. Fig. 31.5: Courtesy Richard Kessel and Randy Kandon, from *Tissues and Organs*, W.H. Freeman. Fig. 31.9: C. Edelmann/Photo Researchers. Fig. 31.13: From A.P. McCauley and J.S. Geller, "Guide to Norplant Counseling", *Population Reports*, Sept. '92, Series K, No. 4, p. 4; photo courtesy Johns Hopkins University Population Information Program. **Chapter 32** Fig. 32.1a: Courtesy Jonathan Van Blerkom, University of Colorado, Boulder. Fig. 32.1b: Doug Perrine/DRK Photo. Fig. 32.2a: G. Shih and R. Kessel/Visuals Unlimited. Fig. 32.2b: Courtesy A.L. Colwin and L.H. Colwin. Fig. 32.3: Courtesy E.M. Eddy and B.M. Shapiro. Fig. 32.4: Courtesy Richard Kessel and Gene Shih. Fig. 32.10b: Courtesy L. Saxen and S. Toivonen. Fig. 32.13b: Courtesy K. Tosney, from *Tissue Interactions and Development* by Norman Wessels. Fig. 32.14: Dwight Kuhn. Fig 32.17: Courtesy Lennart Nilsson, from *A Child Is Born*. **Part 6 Opener:** David Doubilet/National Geographic Society. **Chapter 33** Fig. 33.1: Ivan Polunin/NHPA. Fig. 33.3: Courtesy of Victor A. McKusick, Medical Genetics Department, Johns Hopkins University. Fig. 33.4a: Sharon Cummings/Dembinsky Photo Associates. Fig. 33.4b: © Ron Kimball Studios. Fig. 33.5b-c: Courtesy Professor Lawrence Cook, University of Manchester. Fig. 33.8a: COMSTOCK, Inc. Fig. 33.8b: Tom McHugh/AllStock, Inc. Fig. 33.8c: Kevin Schafer & Martha Hill/Tom Stack & Associates; (inset) Courtesy K.W. Barthel, Museum beim Solenhofer Aktienverein, Germany. Bioline: (Fig. a) Robert Shallenberger; (Fig. b) Scott Camazine/Photo Researchers; (Fig. c) Jane Burton/Bruce Coleman, Inc.; (Fig. d) R. Konig/Jacana/The Image Bank; (Fig. e) Nancy Sefton/Photo Researchers; (Fig. f) Seaphoto Limited/Planet Earth Pictures. Fig. 33.10a: John Garrett/Tony Stone World Wide. Fig. 33.10b: Heather Angel. Fig. 33.11a: Frans Lanting/Minden Pictures, Inc. Fig. 33.11b: S. Nielsen/DRK Photo. Fig. 33.12: Zeisler/AllStock, Inc. Fig. 33.13a: Zig Leszczynski/Animals Animals. Fig. 33.13b: Art Wolfe/AllStock, Inc. Fig. 33.16a: Gary Milburn/Tom Stack & Associates. Fig. 33.16b: Tom McHugh/Photo Researchers. Fig. 33.16c: Nick Bergkessel/Photo Researchers. Fig. 33.16d: C.S. Pollitt/Australasian Nature Transparencies. **Chapter 34** Fig. 34.1a: Leonard Lee Rue III/Earth Scenes/Animals Animals. Fig. 34.1b: Bradley Smith/Earth Scenes/Animals Animals. Fig. 34.4a: Courtesy Professor George Poinar, University of California, Berkeley. Fig. 34.4b: David Muench Photography. Fig. 34.5a: William E. Ferguson. Fig. 34.7a: Courtesy Merlin Tuttle. Fig. 34.7b: Stephen Dalton/Photo Researchers. Fig. 34.9: David Brill. Fig. 34.10: Courtesy Institute of Human Origins. Fig. 34.12: John Reader/Science Photo Library/Photo Researchers. **Chapter 35** Fig. 35.2a: Courtesy Dr. S.M. Awramik, University of California, Santa Barbara. Fig. 35.2b: William E. Ferguson. Fig. 35.4: Kim Taylor/Bruce Coleman, Inc. Fig. 35.5: Courtesy Professor Seilacher. Biuoline: Louie Psihoyos/Matrix. Fig. 35.11 and 35.13 Carl Buell. **Part 7 Opener:** Y. Arthus/Peter Arnold. **Chapter 36** Fig. 36.1: Courtesy Zoological Society of San Diego. Fig. 36.2a-b: David M. Phillips/Visuals Unlimited. Fig. 36.2c: Omikron/Science Source/Photo Researchers. Fig. 36.3a: Courtesy Wellcom Institute for the History of Medicine. Fig. 36.3b: Courtesy Searle Corporation. Fig. 36.5a: A. M. Siegelman/Visuals Unlimited. Fig. 36.5b: Science Vu/Visuals Unlimited. Fig. 36.5c: John D. Cunningham/Visuals Unlimited. Fig. 36.6: (top) Courtesy Dr. Edward J. Bottone, Mount Sinai Hospital, New York; (bottom) Courtesy R.S. Wolfe and J.C. Ensign. Fig. 36.7: CNRI/Science Source/Photo Researchers. Fig. 36.8: Courtesy C.C. Remsen, S.W. Watson, J.N. Waterbury and H.S. Tuper, from *J. Bacteriology*, vol 95, p. 2374, 1968. Human Perspective: Rick Rickman/Duomo Photography, Inc. Fig. 36.9: Sinclair Stammers/Science Source/Photo Researchers. Fig. 36.10: Courtesy R.P. Blakemore and Nancy Blakemore, University of New Hampshire. Bioline: (Fig. 1) Courtesy B. Ben Bohlool, NiftAL; (Fig. 2) Courtesy Communication Arts. Fig. 36.11: (left) Courtesy Dr. Russell, Steere, Advanced Biotechnologies, Inc.; (inset) CNRI/Science Source/Photo Researchers. 36.12b: Richard Feldman/Phototake. **Chapter 37** Fig. 37.1a: Jerome Paulin/Visuals Unlimited. Fig. 37.1b: Stanley Flegler/Visuals Unlimited. Fig. 37.1c: Victor Duran/Sharnoff Photos. Fig. 37.1d: Michael Fogden/DRK Photo. Figure 37.2: R. Kessel and G. Shih/Visuals Unlimited. Fig. 37.3: Courtesy Romano Dallai, Department of Biology, Universitá de Siena. Fig. 37.5a: M. Abbey/Visuals Unlimited. Fig. 37.5b: James Dennis/CNRI/Phototake. Fig. 37.8a: Manfred Kage/Peter Arnold. Fig. 37.8b: Eric Grave/Science Source/Photo Researchers. Fig. 37.9: Mark Conlin. Fig. 37.10: John D. Cunningham/Visuals Unlimited. Fig. 37.11a: David M. Phillips/Visuals Unlimited. Fig. 37.11b: Kevin Schafer/Tom Stack & Associates. Fig. 37.12a: E.R. Degginger. Fig. 37.12b: Waaland/Biological Photo Service. Fig. 37.13: Dr. Jeremy Burgess/Science Source/Photo Researchers. Fig. 37.14: Sylvia Duran Sharnoff. Fig. 37.15: Herb Charles Ohlmeyer/Fran Heyl Associates. Fig. 37.16 and 37.17a: Michael Fogden/Earth Scenes/Animals Animals. Fig. 37.17b: Stephen Dalton/NHPA. **Chapter 38** Fig. 38.1a: (left) Courtesy Dr. C.H. Muller, University of California, Santa Barbara; (inset) William E. Ferguson. Fig. 38.1b: Doug Wechsler/Earth Scenes. Fig. 38.2: T. Kitchin/Tom Stack & Associates. Fig. 38.3a: Michael P. Gadomski/Earth Scenes/Animals Animals. Fig. 38.3b: Courtesy Edward S. Ross, California Academy of Sciences. Fig. 38.3c: Rod Planck/Photo Researchers. Fig. 38.3d: Kurt Coste/Tony Stone World Wide. Bioline: (Fig. a) Michael P. Gadomski/Photo Researchers; (Fig. b-c) E.R. Degginger; (Fig. d) COMSTOCK, Inc. Fig. 38.6a: Kim Taylor/Bruce Coleman, Inc. Fig. 38.6b: Breck Kent. Fig. 38.6c: Flip Nicklin/Minden Pictures, Inc. Fig. 38.6d: D.P. Wilson/Eric & David Hosking/Photo Researchers. Fig. 38.7: J.R. Page/Valan Photos; (inset) Stan E. Elems/Visuals Unlimited. Fig. 38.8a and c: Runk/Schoenberger/Grant Heilman Photography. Fig. 38.8b: John Gerlach/Tom Stack & Associates. Fig. 38.11: John H. Trager/Visuals Unlimited. Fig. 38.13: John D. Cunningham/Visuals Unlimited. Fig. 38.14: Milton Rand/Tom Stack & Associates. Fig. 38.17: Cliff B. Frith/Bruce Coleman, Inc. Fig. 38.18: Leonard Lee Rue/Photo Researchers. Fig. 38.19: Anthony Bannister/Earth Scenes. Fig. 38.21a: J. Carmichael/The Image Bank. Fig. 38.21b: Sebastio Barbosa/The Image Bank. Fig. 38.21c: Holt Studios/Earth Scenes. Fig. 38.21d: Gwen Fidler/COMSTOCK, Inc. **Chapter 39** Fig. 39.1a: Courtesy Karl G. Grell, Universität Tübingen Institut für Biologie. Fig. 39.1b: François Gohier/Photo Researchers. Fig. 39.2a: Larry Ulrich/DRK Photo. Fig. 39.2b: Christopher Newbert/Four by Five/SUPERSTOCK. Fig. 39.6a: Chuck Davis. Fig. 39.7a: Robert & Linda Mitchell. Fig. 39.7b: Runk/Schoenberger/Grant Heilman Photography. Fig. 39.7c: Fred Bavendam/Peter Arnold. Fig. 39.10a: Carl Roessler/Tom Stack & Associates. Fig. 39.10b: Goivaux Communication/Phototake. Fig. 39.10c: Biomedia Associates. Fig. 39.14a: Gary Milburn/Tom Stack & Associates. Fig. 39.14b: E.R. Degginger. Fig. 39.14c: Christopher Newbert/Four by Five/SUPERSTOCK. Fig. 39.15: Courtesy D. Phillips. Fig. 39.16a: Marty Snyderman/Visuals Unlimited. Fig. 39.16b: Robert & Linda Mitchell. Fig. 39.18a: John Cancalosi/Tom Stack & Associates. Fig. 39.18b: John Shaw/Tom Stack & Associates. Fig. 39.18c: Patrick Landman/Gamma Liaison, Fig. 39.20a: Robert & Linda Mitchell. Fig. 39.20b: Edward S. Ross, National Geographic, *The Praying of Predators*, February 1984, p. 277. Fig. 39.20c: Maria Zorn/Animals Animals. Fig. 39.21: Courtesy Philip Callahan. Fig. 39.22a: Biological Photo Service. Fig. 39.22b: Kim Taylor/Bruce Coleman, Inc. Fig. 39.22c: Tom McHugh/Photo Researchers. Fig. 39.23a: E.R. Degginger. Fig. 39.23b: Christopher Newbert/Four by Five/SUPERSTOCK. Fig. 39.26b: Dave Woodward/Tom Stack & Associates. Fig. 39.26c: Terry Ashley/Tom Stack & Associates. Fig. 39.28a: Russ Kinne/COMSTOCK, Inc. Fig. 39.28b: Ken Lucas/Biological Photo Service. Fig. 39.30: Marc Chamberlain/Tony Stone World Wide. Fig. 39.31a: W. Gregory Brown/Animals Animals. Fig. 39.31b: Chris Newbert/Four by Five/SUPERSTOCK. Fig. 39.32a: Zig Leszcynski/Animals Animals. Fig. 39.32b: Chris Mattison/Natural Science Photos. Fig. 39.33a: Michael Fogden/Animals Animals. Fig. 39.33b: Jonathan Blair/Woodfin Camp & Associates. Fig. 39.34a: Bob and Clara Calhoun/Bruce Coleman, Inc. Fig.

39.35a-b: Dave Watts /Tom Stack & Associates. Fig. 39.35c: Boyd Norton. **Part 8 Opener:** Carr Clifton. **Chapter 40** Fig. 40.1: Courtesy Earth Satellite Corporation. Fig. 40.4: Norbert Wu/Tony Stone World Wide. Fig. 40.5: Stephen Frink/AllStock, Inc. Fig. 40.6a-b: Nicholas De Vore III/Bruce Coleman, Inc. Fig. 40.6c: Juergen Schmitt/The Image Bank. Fig. 40.7a: Frans Lanting/Minden Pictures, Inc. Fig. 40.7b: D. Cavagnaro/DRK Photo. Fig. 40.7c: Jeff Gnass Photography. Fig. 40.7d: Larry Ulrich/DRK Photo. Fig. 40.9: Joel Arrington/Visuals Unlimited. Fig. 40.11a: K. Gunnar/Bruce Coleman, Inc. Fig. 40.11b: Claude Rives/Agence C.E.D.R.I. Fig. 40.11c: Maurice Mauritius/Rapho/Gamma Liaison. Human Perspective: Will McIntyre/AllStock, Inc. Fig. 40.13: Jim Zuckerman/West Light. Fig. 40.14: James P. Jackson/Photo Researchers. Fig. 40.15: Steve McCutcheon/Alaska Pictorial Service. Fig. 40.16: Linda Mellman/Bill Ruth/Bruce Coleman, Inc. Fig. 40.17: Bill Bachman/Photo Researchers. Fig. 40.18: Nigel Dennis/NHPA. Fig. 40.19: Bob Daemmrich/The Image Works. Fig. 40.20: B.G. Murray, Jr./Animals Animals. Fig. 40.21a: Breck Kent. Fig. 40.21b: Carr Clifton. **Chapter 41** Fig 41.1a: Gary Braasch/AllStock, Inc. Fig. 41.1b: George Bermard/NHPA. Fig. 41.1c: Stephen Krasemann/NHPA. Fig. 41.2: Robert & Linda stures, Inc. Fig. 41.7a: E.R. Degginger/Photo Researchers. Fig. 41.7b: Anthony Mercieca/Natural Selection. Fig. 41.9: Walter H. Hodge/Peter Arnold. Fig. 41.16: Larry Ulrich/DRK Photo. Fig. 41.18: © Gary Braasch. **Chapter 42** Fig. 42.1a: Andy Callow/NHPA. Fig. 42.1b: Al Grotell. Fig. 42.5a: Cosmos Blank/Photo Researchers. Fig. 42.5b: Karl and Steve Maslowski/Photo Researchers. Fig. 42.5c: Michael Fogden/Bruce Coleman, Inc. Fig. 42.5d: Gregory G. Dimijian, M.D./Photo Researchers. Fig. 42.6: B&C Calhoun/Bruce Coleman , Inc. Fig. 42.7: Kjell Sandved/Photo Researchers. Fig. 42.8a: Wolfgang Bayer/Bruce Coleman, Inc. Fig. 42.8b: J. Carmichael/The Image Bank. Fig. 42.9: Anthony Bannister/NHPA. Fig. 42.10a: Adrian Warren/Ardea London. Fig. 42.10b: Peter Ward/Bruce Coleman, Inc. Fig. 42.10c: Michael Fogden. Fig. 42.11: P.H. & S.L. Ward/Natural Science Photos. Fig 42.12: Tom McHugh/Photo Researchers. Fig. 42.13: E.R. Degginger. Fig. 42.14: William H. Amos. Fig. 42.16: Dr. J.A.L. Cooke/Oxford Scientific Films/Animals Animals. Fig. 42.17: David Cavagnaro/Visuals Unlimited. Fig. 42.19: Stephen G. Maka. Fig. 42.20: Raymond A. Mendez/Animals Animals; (inset): Eric Grave/Phototake. **Chapter 43** Fig. 43.1a: Steve Krasemann/DRK Photo. Fig. 43.1b: Carr Clifton. Fig. 43.1c: Willard Clay. Fig. 43.3: Robert & Linda Mitchell. **Chapter 44** Fig. 44.1: Johnny Johnson/DRK Photo. Fig. 44.2: Dwight R. Kuhn. Fig. 44.3: Mike Severns/Tom Stack & Associates. Fig. 44.4: Bettmann Archive. Fig. 44.5: Robert Maier/Animals Animals. Fig. 44.6: Courtesy Masao Kawai, Primate Research Institute, Kyoto University. Fig. 44.7: Glenn Oliver/Visuals Unlimited. Fig. 44.8: Konrad Wothe/Bruce Coleman, Ltd. Fig. 44.9: Gordon Wiltsie/Bruce Coleman, Inc. Fig. 44.10: Hans Reinhard/Bruce Coleman, Ltd. Fig. 44.11: Oxford Scientific Films/Animals Animals. Fig. 44.12: Patricia Caulfield. Fig. 44.13: Roy P. Fontaine/Photo Researchers. Fig. 44.14: Anup and Manoj Shah/Animals Animals. **Appendix B** Fig. b: Bob Thomason/Tony Stone World Wide. Fig. c: Dr. Gennavo/Science Photo Library/Photo Researchers. Fig. d: F.R. Turner, Indiana University/Biological Photo Service.

Index

Note: A t following a page number denotes a table, f denotes a figure, and n denotes a footnote.

A

C

◄ D ►

◄ E ►

F

I

N